Dialogues Between Artistic Research and Science and Technology Studies

This edited volume maps dialogues between science and technology studies research on the arts and the emerging field of artistic research. The main themes in the book are an advanced understanding of discursivity and reasoning in arts-based research, the methodological relevance of material practices and things, and innovative ways of connecting, staging, and publishing research in art and academia. This book touches on topics including studies of artistic practices; reflexive practitioners at the boundaries between the arts, science, and technology; non-propositional forms of reasoning; unconventional (arts-based) research methods and enhanced modes of presentation and publication.

Henk Borgdorff is Academic Director and Professor of Research in the Arts at the Academy of Creative and Performing Arts, Faculty of Humanities, Leiden University, and Professor at the University of the Arts The Hague, Royal Conservatoire, The Netherlands.

Peter Peters is Professor at the Faculty of Arts and Social Sciences, Maastricht University, The Netherlands.

Trevor Pinch is Goldwin Smith Professor of Science and Technology Studies at Cornell University, USA.

Cover image: The Mutual Wave Machine, an interactive art/neuroscience project by Suzanna Dikker and Matthias Oostrik, image courtesy Lexus Hybrid Art 2014.

Routledge Advances in Art and Visual Studies

This series is our home for innovative research in the fields of art and visual studies. It includes monographs and targeted edited collections that provide new insights into visual culture and art practice, theory, and research.

Film and Modern American Art
The Dialogue between Cinema and Painting
Katherine Manthorne

Bridging Communities through Socially Engaged Art
Edited by Alice Wexler and Vida Sabbaghi

Abstract Painting and the Minimalist Critiques
Robert Mangold, David Novros, and Jo Baer in the 1960s
Matthew L. Levy

Arte Ambientale, Urban Space, and Participatory Art
Martina Tanga

Theory of the Art Object
Paul Crowther

The Digital Interface and New Media Art Installations
Phaedra Shanbaum

Ecocriticism and the Anthropocene in Nineteenth Century Art and Visual Culture
Edited by Emily Gephart and Maura Coughlin

Popularization and Populism in the Visual Arts
Attraction Images
Edited by Anna Schober

Dialogues Between Artistic Research and Science and Technology Studies
Edited by Henk Borgdorff, Peter Peters, and Trevor Pinch

For a full list of titles in this series, please visit https://www.routledge.com/Routledge-Advances-in-Art-and-Visual-Studies/book-series/RAVS

Dialogues Between Artistic Research and Science and Technology Studies

Edited by
Henk Borgdorff, Peter Peters,
and Trevor Pinch

Routledge
Taylor & Francis Group
NEW YORK AND LONDON

First published 2020
by Routledge
52 Vanderbilt Avenue, New York, NY 10017

and by Routledge
2 Park Square, Milton Park, Abingdon, Oxon, OX14 4RN

Routledge is an imprint of the Taylor & Francis Group, an informa business

Library of Congress Cataloging-in-Publication Data
Names: Borgdorff, Henk, editor. | Peters, Peter, 1957– editor. |
Pinch, Trevor, 1952– editor.
Title: Dialogues between artistic research and science and technology studies /
edited by Henk Borgdorff, Peter Peters, and Trevor Pinch.
Description: New York, NY : Routledge, 2020. |
Series: [Routledge advances in art and visual studies] | Includes bibliographical references and index. |
Identifiers: LCCN 2019025791 (print) | LCCN 2019025792 (ebook) |
ISBN 9781138343856 (hardback) | ISBN 9780429438875 (ebook)
Subjects: LCSH: Art—Research. | Research—Methodology. |
Science—Philosophy. | Technology—Philosophy.
Classification: LCC N85 .D53 2050 (print) | LCC N85 (ebook) |
DDC 700.1—dc23
LC record available at https://lccn.loc.gov/2019025791
LC ebook record available at https://lccn.loc.gov/2019025792

ISBN-13: 9781138343856 (hbk)

ISBN-13: 9781032465661 (pbk)

ISBN-13: 9780429438875 (ebk)

Typeset in Sabon
by codeMantra

Contents

Contributors

Ruth Benschop is reader at the Research Centre for Arts, Autonomy and the Public Sphere (Zuyd University of Applied Sciences), Maastricht. She was trained as a theoretical psychologist at Leiden University and finished her PhD (with honours) in the field of science and technology studies at the University of Groningen in 2001. At Maastricht University, she conducted postdoctoral research into sound art and the democratization of music making. Engagement and artistic research are the themes of the center Ruth has led since 2013. She aims to develop artistic research as an experimental, intimate ethnography in which systematic sensitivity for the world is fundamental. The center stimulates work that explores the riches hidden between opposing clichés about art, society, politics, academia, and research.

Henk Borgdorff is Professor of Research in the Arts and Academic Director of the Academy of Creative and Performing Art, Leiden University and professor ('lector') at the Royal Conservatoire, University of the Arts, The Hague. He was editor of the *Journal for Artistic Research* (until 2015), and president of the Society for Artistic Research (until 2019). He has published widely on the theoretical and political rationale of artistic research. A selection is published as *The Conflict of the Faculties: Perspectives on Artistic Research and Academia* (Leiden University Press, 2012). See his profile page on the Research Catalogue: www.researchcatalogue.net.

Claude Draude is professor at the faculty of Electrical Engineering/Computer Science and head of the work group Gender/Diversity in Informatics Systems (GeDIS) at the University of Kassel, Germany. She is on the directors' board of ITeG – Research Center for Information System Design (Kassel, Germany). Her work integrates approaches from feminist STS, New Materialism and arts and design into computing. Her research activities comprise sociotechnical systems and participatory design, human models and artificial intelligence, human–computer interaction, and epistemology in computing – see her book *Computing Bodies: Gender Codes and Anthropomorphic Design at the Human-Computer Interface* (Springer VS, 2017). For more information visit: www.claudedraude.com

Desiree Förster is a PhD student at the Institute for Arts and Media, University of Potsdam, Germany. Currently she is a visiting scholar at the University of Chicago, Department of Cinema and Media Studies. She graduated in philosophy, literature (BA) and media-culture-analysis (MA). Her research interest lies in questions of embodiment, subjectivity, human-animal studies, new materialism, and media ecologies. She was a visiting scholar at the Rensselaer Polytechnique Institut, Troy, NY, in 2017, and holds the Emergent Scholar Award of the New Directions in the Humanities Research Network. As an independent curator, she organizes events at the intersection of art and theory.

Esa Kirkkopelto is a philosopher, artist-researcher and performing artist. He has been working at the University of the Arts Helsinki, as a professor of artistic research (2007–2017) and as the head of the post-doctoral Centre for Artistic Research (CfAR, 2017–2018). Currently, he works there as a visiting researcher. He is the leader of a collective research project Actor's Art in Modern Times on the psychophysical actor training (since 2008), the initiator of the International Platform for Performer Training (since 2014), core-convener or the Performance Philosophy association and the founding member of the Other Spaces live art group (since 2004). His research focuses on the deconstruction of the performing body, both in theory and in practice.

Flora Lysen is a PhD-researcher at the Media Studies Department of the University of Amsterdam. Her research combines approaches from the history of science and media studies to examine how scientists, science educators, and artists use new media to image and imagine the active brain as 'live brain'. Between 2017 and 2019, she was the first program coordinator of the Amsterdam Research Institute of the Arts and Sciences. Before starting her PhD, Flora worked as a cultural programmer, curator, researcher, and teacher for several cultural institutions, including the Haus der Kulturen der Welt in Berlin and the Royal Academy of Art in the Hague.

Peter Peters is professor at the Faculty of Arts and Social Sciences, Maastricht University, and director of the Maastricht Centre for the Innovation of Classical Music. He has published on music, time, mobility, and travel in technological cultures. His current research is concerned with the production of knowledge in artistic practices, and the innovation of musical cultures. Between 2007 and 2013, Peter was professor at the research center Autonomy and the Public Sphere in the Arts (AOK) at the arts faculties of Zuyd University, Maastricht. Here, he developed research lines on art in the public sphere, artistic research, and community arts.

Jon Pigott is an artist, researcher, and academic, whose work explores sound, materials, and systems. His creative practice and research draw on histories and theories of sound art, kinetic sculpture, and technology as well as various making processes including digital fabrication and hand-made electronics. Jon leads the Artist Designer Maker degree program at Cardiff Metropolitan University. Prior to entering academia he worked in the music and audio industry where he helped bring numerous high-profile music and film projects to fruition. Jon was awarded his PhD by Bath Spa University in 2017 and has exhibited and published work internationally both in a solo and collaborative context.

Trevor Pinch, PhD, is Goldwin Smith Professor of Science and Technology Studies at Cornell University. His main research centers on three areas: (1) the sociology of technology and how users engage with technology, (2) sound studies and music and in particular the development of musical instruments and sound objects, (3) markets and the economy with specific attention to the study of selling and persuasion.

His most recent book is a series of interviews conducted by Italian media scholar, Simone Tosini, *Entanglements: Conversations on the Human Traces of Science, Technology and Sound* (MIT Press, 2017). He is also a performing musician with the Electric Golem. He was awarded an honorary degree by the University of Maastricht in 2012.

Hannah Star Rogers is visiting scholar at the University of Edinburgh, where she focuses on the intersection of art and science. She received an MFA from Columbia University and a PhD from Cornell University. In addition to her academic research, Rogers curates art and science exhibits. Recent projects have included Emerge: Artists and Scientists Redesign the Future (Arizona State University), Making Science Visible: The Photography of Berenice Abbott (University of Virginia), and Art's Work in the Age of Biotechnology: Shaping Our Genetic Futures (North Carolina State University).

Johanna Schindler is a cultural scholar and currently a researcher and coordinator in the Department of Literature and Humanities at Haus der Kulturen der Welt, Berlin, where she conceives events that span the arts, sciences, and theory. Previously, she has been a research fellow at Zeppelin University, Germany, from which she also holds a PhD. She has studied cultural studies (BA) and communication and cultural management (MA) and her research interests include cultural production, arts organizations, and epistemic practices in the context of artistic research.

Philippe Sormani is a sociologist working at the intersection of ethnography, ethnomethodology, and science and technology studies. He is also a former head of the academic program at the Swiss Institute in Rome. In 2014, he published a reflexive ethnography of experimental physics, entitled *Respecifying Lab Ethnography*, before coediting *Practicing Art/Science: Experiments in an Emerging Field* in 2018 (both Routledge). Currently, he is based at the STS Lab of the University of Lausanne, and affiliated at the Centre d'étude des mouvements sociaux (CEMS) of the School of Advanced Studies in the Social Sciences (EHESS) in Paris.

Nora S. Vaage is an art historian turned philosopher of art and culture, with a PhD in philosophy of science and ethics. She writes and teaches on a number of topics at the intersection between culture, society and technology. Nora works as assistant professor at Maastricht University and is associate program director of the BA Arts and Culture, an interdisciplinary teaching program combining perspectives on science and technology, politics, media, and the arts. She is particularly interested in how values and knowledge views (ethics and epistemology) affect artistic and scientific work.

Katharina Vones is a lecturer in jewellery and metal design at the University of Dundee. Trained at the University of St Andrews (2001), the Edinburgh College of Art (2006), the Royal College of Art (2010) and the University of Dundee (2017), her research focuses on sustainable craft practice centered on novel materiality and how digital jewellery that possesses biomimetic characteristics can be brought to life through the use of smart materials. Katharina actively blogs about her practice as a way to encourage craft practitioners to participate in open and sustainable communities of making at www.smart-jewellery.com.

1 Dialogues between Artistic Research and Science and Technology Studies

An Introduction

Henk Borgdorff, Peter Peters, and Trevor Pinch

The past two decades have witnessed a new convergence between artistic and scientific ways of knowing and making. Artists not only increasingly draw upon developments in science and technology, but artistic practices are also seen now as the locus of research, presented to and evaluated in art worlds and academia. Scientists are interested in how the arts can contribute to generating new forms of knowledge, methodologies, and engagements. In this book, we aim to explore this convergence from the perspective of two interdisciplinary fields, artistic research and science and technology studies (STS). Artistic research, or research in and through art and design, has gained currency since the 1990s in and beyond higher arts education. Artist-scholars in this field focus on the knowledge, understanding, and experiences enacted in creative processes and embodied in artistic products such as artworks, compositions, and performances. The field of STS has been growing since the 1960s when it was first established by scientists and engineers who were critical of new techniques and developments emerging from science such as genetic engineering, the growing environmental crisis, and the spread and impact of large-scale technological systems such as nuclear power. It now provides a deep understanding of how science and technology work internally, as institutions, and as a body of practices that permeate almost all areas of modern life. In this Introduction, we argue that a dialogue between the two fields can contribute to a reflection on their epistemologies, methodologies, and the ways in which their research outcomes can become public.

STS scholars have studied the arts in relation to questions about science and its history, exploring the role of artists in creating the visual apparatus used by scientists (Jones & Galison, 2014) or the transport of musical notation conventions to the study of sounds and acoustics (Bruyninckx, 2018), to give two examples. Recently, work in STS has focused on the backstage, practical, and preparatory activities constituting works of art or people's engagement with these works (Saaze, 2013). The interest in artistic practices can be linked to research agendas in STS such as subjectivity and the senses; technology and materiality; boundary work; and embodied, situated, and enacted forms of cognition (Benschop, 2009). STS emphasizes the constitutive role of material and social practices in the production of knowledge and technologies. This 'practice turn' is also manifest in the field of artistic research, positioned at the interface of art worlds and academic research. In artistic research, creating performances or artefacts becomes the vehicle in a methodological sense through which knowledge and understanding can be gained. Epistemologically these artefacts and performances embody the knowledge and understanding we gain.

The type of research that we are interested in in this book does not easily fit the conventional frameworks and values of actors and institutions in science and technology as well as in art worlds. One might even argue that the term 'dialogues' in the title is misleading because the convergence between artistic and scientific ways of knowing has been accompanied by controversies (Borgdorff, 2012), some of which will be discussed in this volume.[1] These focus mainly on the demarcation of scientific and artistic practices, their institutions, and the criteria according to which their outcomes are to be valued. For some in the art world, artistic research undermines the modernist dichotomy of autonomy and instrumentalism, breaking away from the alleged 'otherness' of art as a societal domain that has clear boundaries and that can be separated from science (Nowotny, 2010, p. xx). In academia, taking art to be a form of doing research and presenting the works of art that result from that research as a form of knowledge is criticized as conflicting with standards of intersubjectivity, detachment, and justification.

The debate on art as research addresses fundamental philosophical questions of epistemology and methodology and issues of artistic agency and autonomy, as well as institutional and educational strategies. When does art practice count as research? What is the object of artistic research and in what ways is it different from the object of scientific research? How can scientific knowledge be distinguished from knowledge generated within artistic practice? Are scientific research methods radically different from artistic methods of research? In the debates on these questions, one encounters powerful dualisms: art and science, worlds and words, art practice and writing, embodied and discursive knowledge, original artworks and their representations. As a practice, art is often taken to be a paragon of unmethodological, autonomous, and intuitive work, while science appears as methodological, intersubjective, and articulate (Benschop, Peters, & Lemmens, 2014).

Dismantling dualisms and showing how the distinctions they articulate are constructed rather than given belongs to the core strategies of science and technology studies. Transferred to the demarcation debates around art as research, some scholars have followed this strategy by focusing on the sociomaterial practices that bring artworks into being, rather than on their construction as a singular work that can be (re)presented and categorized in a more or less unproblematic way (Latour & Lowe, 2011; Saaze, 2013). A similar genealogical approach that does not take the artwork 'itself' for granted is advocated by Howard Becker, providing insights into how these 'objects and performances take their shape within the daily labour of artists and their collaborators' (Becker, Faulkner, & Kirshenblatt-Gimblett, 2006, p. 13). Following this line of argument, in this book we aim to move beyond the common-knowledge and the self-understandings of science and the arts and instead study and analyze what artist-researchers actually do (Acord & DeNora, 2008; Becker, 2008).

From an STS perspective, it is interesting to explore how distinctions between aesthetic and epistemic outcomes and criteria are crafted by artistic researchers and the respective communities to which they present their work. In addition, artistic research may enrich the methodological repertoire in STS. Artistic researchers in turn, will find much in STS that allows them to reflect in novel ways on their own practices, as Nowotny has argued (2010, p. xxii). In this introductory chapter, we will set the stage for the various dialogues, practices, and experiments at the nexus between artistic research and science and technology studies that are presented in this volume. To do so, we will first focus on the practices, methods, and outcomes of artistic research as an emerging field. We will then ask how research in STS could investigate and inform the work done in artistic research, and how artistic research can inform and enrich STS. Finally, we will argue that STS can provide a meta-perspective on the new 'knowing spaces' (Law, 2017) evolving around the intersection of artistic research practices and science and technology studies.

Artistic Research as Program and Practice

Artistic research gained currency in and beyond higher education and research in the last two decades, yet its genealogy can be traced back to the early modern period. At least in European history, the birth of modern science did not imply a departure from artistry and aesthetics. The inherited unity of truth, goodness, and beauty, however, was broken when the life spheres of science, morality, and art grew apart since the eighteenth century. Institutionally and theoretically, these spheres developed into the relatively autonomous realms and institutes of epistemology and science, ethics and law or religion, and aesthetics and art. But since the days of Leonardo da Vinci those demarcations have also always been accompanied by a feeling of discomfort and anxiety, and every now and then attempts were made to overcome the pain of the dissociations. A history of artistic research will have to uncover in detail what moments in the course of time attest of that desire to bridge the domains or to traverse their boundaries. In philosophical aesthetics important moments were when in eighteenth-century rationalism 'sensuous knowledge' was emancipated from its inferior position to an equal, albeit distinctive footing (cf. Kjørup, 2006) or when in German idealism it was proclaimed that 'all art should become science and all science art; poetry and philosophy should be made one' (Schlegel, 1991, p. 14).

In the twentieth century, the emergence of the artistic research program was anticipated by developments in both academia and the art world. The acknowledgment of know-how (Ryle, 1949) and implicit or tacit knowledge (Polanyi, 1958; 1966) as constitutive for the way we understand and act in the world corrected the focus in epistemology on propositional forms of knowing and understanding: a correction correlating to phenomenology, that would eventually also be taken up by contemporary non-reductive cognitive science (Gibson, 1979; Hutchins, 1995; Ingold, 2000; Newen, De Bruin, & Gallagher, 2018). In the art world, the artistic research program was prepared by a proliferation of art-science encounters and collaborations throughout the twentieth century (cf. Sormani, Carbone, & Gisler, 2018) and by the advance of conceptual art since the 1950s.

An important impetus for the advance of artistic research was the reorganization of higher education, especially the inclusion of art schools and academies in the university system of higher education and research. Starting in the English-speaking world (UK, Canada, Australia (see UK Council for Graduate Education, 1997; Strand, 1998), it reached the European continent in the early twenty-first century. The transformation from vocational training programs to university programs involved the introduction of research in the curricula of art departments, paired with the requirement for research output by faculty, mostly practicing artists, of those departments.[2]

The focus in artistic research is on concrete practices and things – creative processes in the studio, performances, compositions, artworks, installations, artistic interventions. These practices and things not only are the object of study, as in traditional humanities or social science research into the arts, rather their agency and performativity is acknowledged and foregrounded. Artworks and artistic practices do something in the sense that they contribute to our knowledge and understanding of the world. This is in line with what is called the practice turn and the material turn in the sciences and humanities (Schatzki, Knorr-Cetina, & Savigny, 2001). Our changed understanding of what practices and things are has renewed the interest in their ontology. Practices and things speak to us – or speak back to us (see Bal, 2002, p. 61). In an epistemological sense they embody knowledge and understanding, and they are methodologically constitutive in producing knowledge and understanding. These insights are also acknowledged in cultural studies, anthropology, heritage studies (Ingold, 2013), and what is called New Materialism, object oriented ontology, or speculative realism (Barad, 2007, cf. Dolphijn & Van der Tuin, 2013)).

Artistic researchers use a diverse range of methods and tools. This methodological pluralism (Borgdorff, 2012; Hannula, Suoranta, & Vadén, 2014) is widely accepted in the field. Depending on the research topic and the aim of the research, one might use methods and techniques that have their provenance in the humanities or in the social sciences or in technology or in a combination, a triangulation of various methods and tools. That being said, one could distinguish between three aspects that are almost always present when conducting an artistic research project. The first is experimentation (Schwab, 2016). The research takes place through and unfolds in artistic practice, in and through making and performing. That is why it is sometimes referred to as studio-based research. The objective of the artistic experiment is not so much to test something – as in a science or engineering laboratory – but to tell something, to convey content. Testing is all about commensuration and standardization (Pinch, forthcoming), but in telling no appeal needs to be made to commensuration. A second characteristic of artistic research is the involvement and engagement of the person or persons who perform the research. Artistic research is participatory research, and as such it shows kinship with ethnography, where the subject–object divide or the fact–value dichotomy are relativized (Atkinson, Coffey, Delamont, Lofland, & Lofland, 2007; Pink, Hubbard, O'Neill, & Radley, 2010). A third feature of artistic research is that the research findings need a form of analysis or interpretation. Here, 'theory' might help to contextualize the research and to show how it relates to other research and how it is embedded in academic, cultural, social, or political spheres and discourses. Artistic research thus appropriates a wide variety of research methods and techniques from other research fields, and it is distinctive in the combination of experimentation, participation, and interpretation.

To demarcate artistic research from other types of research it is generally agreed in the field that artworks, varying from concrete, material artefacts to ephemeral performances or artist interventions, should be part of the outcome of the investigation. The material outcome of the research, however, is not the research itself. Even the documentation of the research outcome, varying from audio or video registrations of performances to exhibition catalogues and so-called 'artist-books,' does not suffice as an account of the research. Additional work has to be done to articulate and communicate the research, to show that it involves 'a process of investigation leading to new insights, effectively shared' (Research Excellence Framework, 2011, p. 48).

In the debate on artistic research, many have taken the position that this additional work is to be seen as the reflective, discursive, or written part of the research or of the submission of a PhD thesis. Hence, there is a sharp distinction between the artwork and the reflection on it.[3] But that position misses the point of the intertwinement of theory and practice in artistic research. If we acknowledge the agency of material practices and things, and if we stress the importance of studio-based, practice-based methods, and if we furthermore acknowledge that cognition is embodied, embedded, and enacted in material practices, then we should not hesitate to conclude that the reasoning is also located in those material practices. One should at least take the agency: that is, the epistemic and methodological force of the artefacts and artistic practices into account, something that is also acknowledged in STS.

How to articulate this style of reasoning? How to articulate the epistemic and methodological force of art? Here we want to underline the role of rich-media articulation, documentation, publication, and dissemination. This is a form of articulation – of writing, one could say – in which artistic material and its documentation is interwoven with text-based material. One of the tasks now is to rethink what 'discursivity' means, what it is to make a claim in and through art, what reasoning is, once we have accepted that material practices and things in this field of inquiry are not only constitutive in a methodological sense but that they also count as valid expressions of research processes and outcomes. Questions such as these have also been taken up in the field of science and technology studies.

An STS Perspective on Artistic Research

Science and technology studies originated in the 1960s in the critical debates on the societal role and impact of scientific research and technological innovations. This criticism was informed by the debate over the role of the social in the history and philosophy of science. Proponents of internalism, typically philosophers such as Karl Popper (1963), claimed that scientific knowledge production is relatively independent of the social, whereas historians such as Thomas Kuhn (1962) argued that the history and dynamics of science cannot be described and understood without taking social factors into account. This debate over the role of the social resonated with the distinction that Popper made between the context of scientific discovery and the context of justification. Kuhn, Paul Feyerabend (2010), and others showed that even the justification of science is co-dependent on contingent factors sometimes from outside the realm of science.

The Sociology of Scientific Knowledge (SSK) and especially the 'Strong Programme' of the Edinburgh School in the 1970s and 1980s pushed the place of social explanation further by seeking to explain how both false and true knowledge claims are socially shaped. In the same period, scholars from the Bath School and its 'Empirical Programme of Relativism' (EPOR) focused more closely on the concrete, material work scientists engage in through ethnographic studies, studies of scientific controversies, and of science-in-the-making (Bloor, 1976; Collins, 1985). Those studies reached beyond or behind the formal reports and protocols of science and focused on the often implicit, tacit knowledge and know-how and the embodied skills that feed into the research processes and marked the outcomes. A landmark study was the ethnographic research that Bruno Latour and Steve Woolgar conducted in a scientific laboratory in the late 1970s, where they followed the everyday work of scientists constructing scientific facts (Latour & Woolgar, 1979).

In the early 1980s, the symmetry principle of the sociology of scientific knowledge – explaining both false and true knowledge claims from social factors – was introduced in research on why some technological innovations were successful and others not. Drawing on the history of the bicycle, Pinch and Bijker (1984) showed that as an artefact the bicycle was interpreted in different ways by different social groups until one interpretation of the bicycle stabilized. Their Social Construction of Technology (SCOT) program aims to understand which cultural, economic, social, and political factors co-determine the course of technological developments. Subsequent research on the social shaping of technology has focused on issues such as the non-linearity of technological innovations, the role of users and non-users in these innovation trajectories, and the ways new technologies are shaped by path-dependency and obduracies (MacKenzie & Wajcman, 1999; Oudshoorn & Pinch, 2003).

In the 1990s, the scholarly debate in STS focused partly on the criticism that too much explanatory force was given to human and social factors, as if the internal logic and dynamics of science can be understood by looking at the intended and unintended actions and interpretations of people alone. Proponents of actor–network theory (ANT) argued for a 'principle of generalized symmetry': to understand scientific, technological, or other practices, we should depart from a priori dualisms between the social and the material or between culture and nature (Latour, 2005). Instead, we should develop a relational account of practices as heterogeneous assemblages of people and their ideas and skills, social institutions and organizations, as well as things such as technical objects, materialities, and apparatuses. Instead of being presented as explanation, the social itself is seen to be constituted, staged, or assembled through the interplay between human and non-human actors. This implies an 'ontological multiplicity': reality is not one thing, nor is it given, but it is constructed, staged, and performed and contingent on how human and non-human actors interact (Mol, 2002).

From the early days of ethnographic laboratory studies, following the actors has been a key research strategy in STS. As an empirical enterprise, it seeks to unravel the dynamics of science and technology-in-the-making, by studying practices. Another characteristic of STS methodologies is a focus on case studies, ranging from bicycles to automated subway trains, bridges, contraceptives, air pumps, and bush pumps.[4] Stabilization of these artefacts and innovations in networks and practices cannot be explained only from their intrinsic properties or qualities but should take into account local circumstances and contingencies. STS case studies often share the core argument that things could have been otherwise. This motto reflects the critical origins of the field, arguing against technological determinism and its agenda of democratizing science and technology by making their development more inclusive and reflexive. Showing how science and technology are socially shaped in their making and use enabled STS researcher to locate and rethink normativities and politics as they emerge in practices. Recently, this study of politics in action has been expanded to the normativity of artistic practices to understand how aesthetic judgments are made (cf. Peters, this volume).

How could STS research inform and inspire the work done in artistic research? To begin with, drawing on its science research, STS can help to analyze how artistic research establishes itself as an emerging field and how knowledge claims are made in this field. Furthermore, its focus on the study of practices as sociomaterial assemblages fits well with the interest in practices in artistic research, as well as the materialities, embodied skills, and sensory knowledge that play an important role in these practices. In addition, the sensitivity for how methods shape the realities that they aim to research resonates with the performative force of studio-based methods in artistic research. And finally, STS shows an interest in how the epistemic and methodological force of art can be articulated and made public through rich-media documentation, publication, and dissemination.

Artistic Research, STS, and Their Knowing Spaces

After our concise and admittedly sketchy overviews of the two fields that this book intends to bring into dialogue, we want to elaborate on the question why and how such a dialogue can be fruitful for both fields. We will map some of the common ground to be found at the level of knowledge production, research methods, and outcomes. We will then argue that the intersections of artistic research practices and science and technology studies can be thought of as new 'knowing spaces' (Law, 2017).

To create works of art or performances, artists have always reworked and adapted existing art, mobilized contexts and sources relevant to their art-making, and developed new skills and technologies. The field of artistic research, however, has made this work more explicit as a form of research that entails knowledge claims. Artistic researchers not only present their art as works or practices that are acknowledged and evaluated in art worlds, they also stage the research that their art-making requires and implies in ways that allow academic communities or other relevant communities to assess its epistemic value. They thus expand the ways in which their artworks and artistic practices can exist and be made relevant. An encounter between artistic research and STS involves asking what kind of knowledge artistic research produces and how its knowledge claims relate to traditional scientific ways of knowing. As Salter, Burri, and Dumit (2017) have argued, art and design as knowledge practices highlight the role of improvisation, creativity, and invention. These practices put embodied knowledge center stage, as well as material engagement and forms of sensory perception. With STS, they share a keen interest in performance and performativity, as well as in the agency situated in material artefacts. Finally, artistic research as knowledge practice is characterized by an interventionist approach that stages different forms of engagement and critique. All of this resonates with work in STS on situated knowledges and situated action where knowing, doing, and making as cognitive and perceptual, embodied and sensory, as well as materially mediated activities are intimately related (Suchman, 2007).

A central insight in STS is that research methods do not only observe and represent materials, issues, and events but in fact act upon and intervene in these materials, issues, and events. Research not only analyzes, documents, and informs, but also performs realities and ontologies and reforms and transforms them through the act of researching (Law, 2004). Changing conceptions of what constitutes the empirical also led to an intensification of interest in research methods as ways of making knowledge in social and cultural research (Lury & Wakeford, 2012). STS and artistic research share the project of enlarging their methodological repertoires, as well as the reflection on the politics of what Law and Ruppert have called 'the material heterogeneities of knowing' (2016, p. 20). Ethnography is an example of how social sciences and artistic practice can share a research method, that through its use in these two different contexts can acquire new sensitivities (Foster, 1996). Artists created situations in which the familiar and the foreign vacillate. Precisely the mechanisms that determine what we take for granted and what we experience as strange thus become the medium of the artist as ethnographer. It is through their public staging of everyday reality in experimental situations that audiences can look at themselves as anthropologists (Schneider & Wright, 2006).

The need for hybrid forms of publication and dissemination that do justice to the non-verbal, non-propositional nature of research outcomes is felt both in artistic research and STS. Extended and intermedial publications not only reflect the hybridity of the research and its methods but also of the publics that are addressed. For example, one of the more vexing topics in the debate on the institutionalization of artistic research in academia are the criteria for an artistic research PhD. In universities that allow artistic researchers to defend their research, as a rule a written text next to an artistic product is requested. This requirement shows how the dichotomies between art and academia continue to exist in practice. Reflection on dissemination strategies and formats to make research public is shared with the field of STS. Here, scholars seek ways to communicate their research results also to wider audiences than can be reached through written scientific work as a contribution to the democratization of science and technology (Marres, Guggenheim, & Wilkie, 2018).

John Law has argued that 'knowing and its methods are materially complex and performative webs of practice that imply particular arrays of subjects, objects, expressions or representations, imaginaries, metaphysical assumptions, normativities, and institutions' (Law, 2017, p. 47). He thinks of these heterogeneous arrays as 'knowing spaces' that can have power and obduracy (Law, 2011) Giving the example of academic knowing spaces, Law explains how access to these spaces depends on the ability and willingness to conform to its conventions, procedures, competences, topics, theoretical frameworks, and criteria. He also gives recent examples of unconventional or hybrid knowing spaces that worked through exhibitions, writing poetry, simulations, reciprocal human–animal interactions, art–science interactions, or activism and participatory methods (Law, 2017, p. 48). In his use of the concept of knowing spaces, and by acknowledging that creating different knowing spaces can be slow, hazardous, uncertain and lonely (ibid.), Law applies a typical STS line of argumentation to practices of knowing, their methods, as well as the reception and workings of their outcomes. What is learned from other case studies in STS is valid here as well:

> That methods are shaped by the social; that they also shape, stage, and structure the social; that they are performative and heterogeneously enact objects, worlds, and realities; that they are situated, productive, essentially political, and normative; and that they might be otherwise. (Law, 2017, p. 48)

Dialogues, Practices, Experiments

The chapters in this book all relate to and reflect on the hybrid knowing spaces at the intersections between artistic research and STS. In our ordering of the chapters, we have placed them under three different headings: 'Dialogues,' 'Practices,' and 'Experiments.' The chapters in the first part of the book all discuss general issues and questions around the encounter between artistic research and STS. They thus contribute to the meta-reflexive debate that accompanies this dialogue. The second part of the book focuses on concrete examples of practices of artistic research, and how these practices can be analyzed using STS concepts and methods.[5] The chapters give detailed accounts of these practices, answering the question of what artistic researchers actually do, either by following their work as academic scholars or by recounting their own practices as artist-researcher. The third part of the book is labelled 'Experiments'. The chapters in this section revolve around one of the central affinities between scientific and artistic research: setting up experimental situations that enable the emergence of knowledge and understanding.

Dialogues

Artworks and artistic practices are meaningful in the art world, whereas they also embody or enact knowledge and insights that function as commodities in academia. In his chapter, Henk Borgdorff approaches this problem of demarcation without reproducing conceptual dichotomies by focusing on what happens when artworks and artistic practices 'travel' from the art world to academia, from the realm of the aesthetic to the realm of the epistemic. What kinds of translations, transformations, or transpositions happen here? Borgdorff answers this question by discussing the process of establishing the online Research Catalogue that functions as a platform for the archiving, documentation, management, publication, and dissemination of artistic research.

The theme of translation is also addressed in the chapter by Esa Kirkkopelto. He argues that the relation between STS and artistic research invites a rethinking of procedures of translation. One of the basic operations in science that depends on translation is measurement. In scientific research this implies that the things under study, which are not necessarily human, are made to speak to us, humans and researchers, so that we can understand them. From the artistic research perspective, however, although things and materials speak to us we cannot necessarily understand their talk, let alone translate it into discursive language. Kirkkopelto poses the questions of how, according to STS, objects are constituted in science, how they are constituted in the arts, and how these processes and their results are similar or different.

Whereas the two previous chapters seem to take the ideal-typical character of art and academia as a starting point to reflect on their interrelations, Ruth Benschop puts this dichotomy aside. Her interest is not so much in defining and defending what artistic research may be, as it is in what artistic researchers do and what the good words are to speak about what they do. Her approach is to conduct a thought experiment on the *craft* of artistic research. This thought experiment consists in deliberately misreading or misplacing two examples, both on the brink of art and ethnography, 'as if' they were artistic research. She reads the work of the ethnographer Stefan Hirschauer like an artist, whereas she understands the interventions of the artist Pilvi Takala as an anthropologist or sociologist. Together, both examples suggest non-reductive ways in which we can grasp both the strictness of emerging methods as well as the space for that which escapes such methods, in academic as well as in artistic work.

How artistic research produces knowledge is the topic of a discourse that has accompanied the field from its beginnings. Drawing on theories from STS and philosophy, Nora Vaage takes bioart practices as a starting point for a meta-reflection on the concept of knowledge itself. In recent years, an increasing number of artists are engaging with the biotechnosciences, entering the laboratories to create art *in vivo*. In what sense of the word can we speak about artistic lab practices as producing knowledge? Whereas a common definition of knowledge in epistemology is justified true belief, this definition reduces the role of art to science communication. Vaage argues that a more suitable concept to apply to the meaning-making of art may be wisdom. Considering artistic research as a practice that aims for wisdom might help create a space for such research that is connected to and complements other academic practices, without having to aim for the same forms of knowledge outcomes.

In her chapter, Hannah Rogers argues that science-and-technology-engaged artists are practicing STS by material means. They share STS's concerns: who gets to set the agenda of science and participate in its workings, and how does science create and maintain its knowledge corpus and related power structures? To see these artists, who are engaged with science and technology, as outside the STS community is a form of boundary-making. Art and science and technology studies (ASTS) is beginning to unpack some of this work and its consequences for STS. Rogers also considers that particular group of contemporary artists, known as bioartists, in order to examine the specific possibilities for overlaps between STS scholarship and artists who are engaged with science and technology. Works like those at SymbioticA, a wet research lab at the University of Western Australia (UWA), should be considered STS by other means: that is, these works engage some of the same issues that science studies engages but do so not by publishing papers but by vesting their ideas into physical and tactical objects.

Practices

Jon Pigott explores how sensibilities and approaches from science and technology studies can help to understand and identify the practice of kinetic sound art. He does so by developing the idea of the 'material system' identified in the work of STS scholars Bruno Latour and John Law and relating it to the object- and material-based technological systems of kinetic sound art. Following the lineage of technologically engaged art practice offers opportunities for an STS of the arts. A first-hand case study of an original kinetic sound piece by the author titled *Electromagnetic Interrogations* (2011–2014) allows further reflection on the artistic construction of technology as well as a consideration of how the making of a technological artwork and the exploration of related STS influenced ideas can be thought of as a single 'method assemblage' (Law, 2004, p. 13). Pigott argues that kinetic sound art often aims to evoke an alternative view of technology as a contingent and evolving system. For these sound artists there is also a tension between communicating this contingent nature of technology and producing technological artworks that will reliably 'perform' and work in the gallery or concert hall. This allows for a reflection on the assemblage nature of methods for simultaneously making connections, insights and artworks.

Johanna Schindler examines in her chapter two collaborative artistic research projects in Germany and Switzerland through ethnographic field research. Interested in the epistemic potential of boundary objects, she focuses on a newly developed digital musical instrument and a computer- and biofeedback-controlled space. The researchers in the projects stemmed from various disciplines such as computer science, musicology, product design, media studies, and media arts. The observed researchers deployed an artistic working mode to create multifunctional objects, which served both as investigative instruments for their research endeavor and as presentation of first results. Even though artistically designed, these objects were neither intended nor considered to be artworks. Rather, they remained works-in-progress and were a first step in the search for an adequate presentation format for the research results. Seeing these objects as boundary objects allows Schindler to show how their design and functionality reflect the researchers' individual backgrounds and research interests and how they structure and re-organize the ongoing research process.

What does lived experience mean in times of environmental crisis? The first-person perspective of the lived body, which in phenomenology is foundational to sensual perception and knowledge creation, seems to be unable to grasp processes on the planetary scale such as climate change, Desiree Förster claims in her chapter. Given the fact that the environmental crisis is so extensive and neither temporally nor spatially understandable to an individual, scholars have called for a new environmental aesthetic. Such a re-situating of human agency into its natural environment combines several fields in their shared ways of re-thinking subjectivity by emphasizing the role non-human powers and processes play on various levels of life and sense-making. Using concepts from phenomenology, New Materialism, and actor–network theory, Förster explores how new aesthetic practices at the intersection of art and design develop forms of incorporating non-human agencies into the lived and sensual experience or expand the human body towards its animated, vital environment.

Recently, musical practices and their technologies have become a research subject in STS as well as in the related field of sound studies. In his chapter, Peter Peters enters the pipe-organ builder's workshop to study ethnographically how materialities, such as metal, wood, and leather, and skills, such as metal casting and pipe voicing, are made to matter artistically. Organs are considered as aesthetic and technological mirrors of their time, which makes the practices of knowing, making, and performing that revolve around them a strategic research site to explore interrelations of the epistemic and the aesthetic. Peters followed the building of a new Baroque organ in the Orgelpark, a venue in Amsterdam that aims to give the pipe organ a new role in musical life. Through his ethnographical observations, he describes how acquiring historical knowledge of organ-building practices and relearning eighteenth-century artisanal skills enabled the organ builders to create a technical space in which to articulate intellectual, tactile, sensory, or aesthetic reasons for the normative claim that a pipe sounds good.

Experiments

Screens are everywhere. Claude Draude addresses this pervasiveness of computer screens by following a phenomenological conception of screens as mattering only as screens-in-the-world. She discusses characteristics of the computer screen, interweaving basic principles of computing and cultural impact. This discussion and the phenomenological conception of screens provide the basis for experimenting with screenness through art-based research. The focus here is not on the artistic product or object as such but on thinking, reflecting, and perceiving through art-based experimental set-ups, with a special interest in embodiment and site-specific situatedness. The quality of the art-based approach, Draude argues, lies in its power to produce an experimental field of non-standard ways of knowledge production in a technological field. Reviewing her art-based experiments leads Draude to shift the focus for future research from the metaphors derived from optics towards the notion of the screen as a membrane. Thus, the screen's own agency as semi-permeable threshold, as well as its interconnectivity to specific sites, bodies, and contexts can be addressed.

Katherina Vones examines the way in which the ancient practice of alchemy and the figure of the alchemist could be used to offer researchers and practitioners, operating at the boundaries between creative and scientific practice, a model for engaging with the concept of cross-disciplinary knowledge generation. Alchemical practice has been connected to craft practitioners, and in particular goldsmiths and jewelers, from the early modern period onwards. It experienced a resurgence during the New Jewellery movement in the 1970s, where the altered perception of material preciousness in jewelry prompted some craft practitioners to return to the conceptual paradigms of alchemy in order to define their practice. More recently, Vones finds, the term 'alchemical craft' has been used to describe practices and practitioners who work with novel materials and processes that have been sourced from the laboratories of researchers, often through interdisciplinary collaborative projects supported by an institutional framework. Materials libraries act as modern day alchemical laboratories, where interested artistic practitioners, makers, materials scientists, and academic researchers gather to experience and discuss novel materiality. Thus a tradition of spaces for experimentation such as those secretive but well-documented meetings that took place between like-minded alchemists in sixteenth-century Europe is revived at a time when such interdisciplinary collaborations are encouraged.

Over the past decades, measurements of the brain's electrical activity have moved beyond the neuroscientific laboratory into other domains, including practices of mindfulness training and meditation, hacker spaces, consumer research, the game industry, and also a variety of art-meets-science events. Flora Lysen investigates these art-science works as public experiments: that is, as configurations of unfinished knowledge, developed with participants as they engage with an art-science installation. She argues that such art-science installations, as entangled experiments, can help to reimagine the empirical and conceptual outlines of research into the social brain but that such intra-disciplinary, ontological reflections are also always paired with other logics, including an assumed critical role of art vis-à-vis science, as well as the position of art in potentially stimulating an innovation-oriented neuro-techno-scientific society. Lysen claims that art-science collaborations, most notably those in the field of bioart, often take the form of art that critically elucidates or examines scientific practices. In this line of reasoning, however, art-science collaborations are intermediating between the fields of art and science, yet they rarely constitute genuine artistic research in the sense of a real hybridization of domains.

In his chapter, Philippe Sormani goes back to the ethno-methodological breaching experiments that Harold Garfinkel developed in the 1960s for a methodological reflection on re-enactment as a research strategy. Sormani explores the interplay between performance art and video analysis. More specifically, his chapter revisits a particular position in performance art – Andrea Fraser's institutional critique *qua* filmed intervention – in dialogue with practice-based video analysis, a recent development in ethno-methodology practiced alongside mainstream STS. Sormani examines three media announcements from a developing corpus of video recordings, all of which announce one form or other of 'machine intelligence,' relating to video gaming, neuromorphic computing, and machine learning, respectively. In his contribution to a 'sociology of demonstrations' Sormani draws together insights from the presented analysis at the tricky intersection of performance art, Fraser's institutional critique, and video analysis, if not contemporary art and current STS more broadly.

Notes

1 The central controversy around artistic research involves its legitimacy as a proper academic field of investigation: that is, whether it conforms to the prevailing standards of scientific research with regard to methodology, replicability, reliability, reporting, and so on. In that controversy some people tend to take sides in such a way that a caricature is made of the opponent. Science is curtailed and reduced to a 'scientistic' picture, where everything that falls outside the scope of the controlled experiment is dismissed as pseudo-science or

fraud. Others see art as the realm where autonomy and resistance towards standards and restrictions prevail. It is our assumption that such an opposition is not helpful when one wants to understand the rationale and internal dynamics of the artistic research program.

2 Academic drift is not a new phenomenon in higher education. The history of universities shows a frequent adaptation to changing circumstances and the inclusion of more and more areas or ways of investigation, starting with the advance of experimental science itself in the seventeenth century, over the breakdown of natural history and philosophy into the sciences and the rise of and the controversies around social science in the nineteenth century, up until the inclusion of technology and design programs and the unrestrained expansion of academia into all kinds of areas in the twentieth century.

3 One can find such a distinction, for instance, in the regulations for the new artistic doctorate in Sweden, Norway, and Austria.

4 The examples of these canonical 'object lessons' are taken from Bijker (1995), Latour (1992), Winner (1999), Shapin and Schaffer (1985), and De Laet and Mol (2000). The list of examples could easily be expanded.

5 For a comparison of artistic practices and STS practices in the world of sound, see Pinch, 2016.

References

Acord, S. K., & DeNora, T. (2008). Culture and the Arts: From Art Worlds to Arts-in-Action. *Annals of the American Academy of Political and Social Science, 619*(1), 223–237.

Atkinson, P., Coffey, A., Delamont, S., Lofland, J., & Lofland, L. (Eds.) (2007). *Handbook of Ethnography.* Thousand Oaks: Sage Publications.

Bal, M. (2002). *Travelling Concepts in the Humanities: A Rough Guide.* Toronto: University of Toronto Press.

Barad, K. (2007). *Meeting the Universe Halfway: Quantum Physics and the Entanglement of Matter and Meaning.* Durham, NC: Duke University Press.

Becker, H. S. (2008). *Art Worlds.* Berkeley: University of California Press.

Becker, H. S., Faulkner, R. R., & Kirshenblatt-Gimblett, B. (Eds.) (2006). *Art From Start to Finish: Jazz, Painting, Writing, and Other Improvisations.* Chicago: University of Chicago Press.

Benschop, R. (2009). STS on Art and the Art of STS: An Introduction. *Krisis: Journal for Contemporary Philosophy, 2009*(1), 1–5.

Benschop, R., Peters, P., & Lemmens, B. (2014). Artistic Researching: Expositions as Matters of Concern. In M. Schwab & H. Borgdorff (Eds.), *The Exposition of Artistic Research: Publishing Art in Academia* (pp. 34–51). Leiden: Leiden University Press.

Bijker, W. E. (1995). *Of Bicycles, Bakelites, and Bulbs: Towards a Theory of Sociotechnical Change.* Cambridge, MA, and London: MIT Press.

Bloor, D. (1976). *Knowledge and Social Imagery.* London: Routledge.

Borgdorff, H. (2012). *The Conflict of the Faculties: Perspectives on Artistic Research and Academia.* Leiden: Leiden University Press.

Bruyninckx, J. (2018). *Listening in the Field: Recording and the Science of Birdsong.* Cambridge, MA: MIT Press.

Collins, H. M. (1985). *Changing Order: Replication and Induction in Scientific Practice.* London: Sage.

De Laet, M., & Mol, A. (2000). The Zimbabwe Bush Pump: Mechanics of a Fluid Technology. *Social Studies of Science, 30*(2), 225–263.

Dolphijn, R., & Van der Tuin, I. (2013). *New Materialism: Interviews and Cartographies.* Ann Arbor: Open Humanities Press.

Feyerabend, P. (2010). *Against Method* [1975], 4th ed. New York: Verso Books.

Foster, H. (1996). *The Return of the Real: The Avant-Garde at the End of the Century.* Cambridge, MA: MIT Press.

Gibson, J. J. (1979). *The Ecological Approach to Visual Perception.* Boston, MA: Houghton Mifflin.

Hannula, M., Suoranta, J., & Vadén, T. (2014). *Artistic Research: Methodology.* Bern: Peter Lang.

Hutchins, E. (1995). *Cognition in the Wild.* Cambridge, MA: MIT press.

Ingold, T. (2000). *The Perception of the Environment: Essays on Livelihood, Dwelling and Skill.* New York: Routledge.

Ingold, T. (2013). *Making: Anthropology, Archaeology, Art and Architecture.* London and New York: Routledge.

Jones, C. A., & Galison, P. (2014). *Picturing Science, Producing Art.* London and New York: Routledge.

Kjørup, S. (2006). *Another Way of Knowing.* Bergen: Bergen National Academy of the Arts.

Kuhn, T. S. (1962). *The Structure of Scientific Revolutions.* Chicago: University of Chicago Press.

Latour, B. (1992). *Aramis, ou l'amour des techniques.* Paris: Éditions la Découverte.

Latour, B. (2005). *Reassembling the Social: An Introduction to Actor-Network-Theory.* Oxford: Oxford University Press.

Latour, B., & Lowe, A. (2011). The Migration of the Aura, or How to Explore the Original Through Its Facsimiles. In T. Bartscherer & R. Coover (Eds.), *Switching Codes: Thinking Through Digital Technology in the Humanities and the Arts* (pp. 275–297). Chicago: University of Chicago Press.

Latour, B., & Woolgar, S. (1979). *Laboratory Life: The Social Construction of the Scientific Facts.* Beverly Hills: Sage Publications.

Law, J. (2004). *After Method: Mess in Social Science Research.* London: Routledge.

Law, J. (2011). The Explanatory Burden: An Essay on Hugh Raffles' Insectopedia. *Cultural Anthropology 26*(3), 485–510.

Law, J. (2017). STS as Method. In U. Felt, R. Fouché, C. A. Miller, & L. M. Smith-Doerr (Eds.), *The Handbook of Science and Technology Studies* (pp. 31–57). Cambridge, MA: MIT Press.

Law, J., & Ruppert, E. (Eds.) (2016). *Modes of Knowing: Resources from the Baroque.* Manchester: Mattering Press.

Lury, C., & Wakeford, N. (2012). *Inventive Methods: The Happening of the Social.* London: Routledge.

MacKenzie, D., & Wajcman, J. (1999). *The Social Shaping of Technology*, 2nd ed. Buckingham: Open University Press.

Marres, N., Guggenheim, M., & Wilkie, A. (2018). *Inventing the Social.* Manchester: Mattering Press.

Mol, A. (2002). *The Body Multiple.* Durham, NC: Duke University Press.

Newen, A., De Bruin, L., & Gallagher, S. (2018). *The Oxford Handbook of 4E Cognition.* Oxford: Oxford University Press.

Nowotny, H. (2010). Foreword. In M. Biggs & H. Karlsson (Eds.), *The Routledge Companion to Research in the Arts* (pp. xvii–xxvi). London: Routledge.

Oudshoorn, N. E. J., & Pinch, T. (Eds.) (2003). *How Users Matter. The Co-construction of Users and Technology.* Cambridge, MA: MIT Press

Pinch, T. J. (2016). 'Bring on Sector Two!' The Sounds of Bent and Broken Circuits. *Sound Studies, 2*(1), 36–51.

Pinch, T. J. (forthcoming). Coda: Testing and Why it Matters. In A. Hiu, M. Mills, and V. Tkaczyk (Eds.), *Testing Hearing: Music, Science, Industry.* Oxford: Oxford University Press.

Pinch, T. J., & Bijker, W. E. (1984). The Social Construction of Facts and Artefacts: Or How the Sociology of Science and the Sociology of Technology Might Benefit Each Other. *Social Studies of Science, 14*(3), 399–441.

Pink, S., Hubbard, P., O'Neill, M., & Radley, A. (2010). Walking across Disciplines: From Ethnography to Arts Practice. *Visual Studies*, *25*(1), 1–7.
Polanyi, M. (1958). *Personal Knowledge: Towards a Post-Critical Philosophy*. London: Routledge and Kegan Paul.
Polanyi, M. (1966). *The Tacit Dimension*. London: Routledge and Kegan Paul.
Popper, K. (1963). *Conjectures and Refutations: The Growth of Scientific Knowledge*. London: Routledge.
Research Excellence Framework. (2014). Assessment Framework and Guidance on Submissions. Retrieved from https://webarchive.nationalarchives.gov.uk/20170302114208/http://www.ref.ac.uk/pubs/2011-02.
Ryle, G. (1949). *The Concept of Mind*. Chicago: University of Chicago Press.
Saaze, V. van. (2013). *Installation Art and the Museum: Presentation and Conservation of Changing Artworks*. Amsterdam: Amsterdam University Press.
Salter, C., Burri, R. V., & Dumit, J. (2017). Art, Design, and Performance. In U. Felt, R. Fouché, C. A. Miller, & L. Smith-Doerr (Eds.), *The Handbook of Science and Technology Studies*, 4th ed. (pp. 139–168). Cambridge, MA: MIT Press.
Schatzki, T. R., Knorr-Cetina, K., & Savigny, E. V. (2001). *The Practice Turn in Contemporary Theory*. New York: Routledge.
Schlegel, F. (1991). *Critical Fragments*, trans. Peter Firchow. Minnesota: University of Minnesota Press.
Schneider, A., & Wright, C. (2006). *Contemporary Art and Anthropology*. Oxford: Berg.
Schwab, M. (Ed.) (2016). *Experimental Systems: Future Knowledge in Artistic Research*. Leuven: Leuven University Press.
Shapin, S., & Schaffer, S. (1985). *Leviathan and the Air-Pump: Hobbes, Boyle and the Experimental Life*. Princeton, NJ: Princeton University Press.
Sormani, Ph., Carbone, G. & Priska, G. (2018). *Practicing Art/Science: Experiments in a Emerging Field*. London: Routledge.
Strand, D. (1998). Research in the Creative Arts. Canberra: Department of Employment, Education, Training and Youth Affairs – Evaluations and Investigations. Program Report 98/6. Retrieved from http://pandora.nla.gov.au/pan/24685/20020426-0000/www.detya.gov.au/archive/highered/eippubs/eip98-6/eip98-6.pdf.
Suchman, L. A. (2007). *Human–Machine Reconfigurations. Plans and Situated Actions*, 2nd ed. Cambridge: Cambridge University Press.
UK Council for Graduate Education. (1997). Practice-Based Doctorates in the Creative and Performing Arts and Design. Retrieved from http://www.ukcge.ac.uk/media/download.aspx?MediaId=1289.
Winner, L. (1999). Do Artifacts Have Politics? In D. MacKenzie & J. Wajcman (Eds.), *The Social Shaping of Technology*, 2nd ed. (pp. 28–40). Buckingham: Open University Press.

Part I

Dialogues

2 Cataloguing Artistic Research

The Passage from Documented Work to Published Research

Henk Borgdorff

Artistic research is a rewarding subject for science and technology studies (STS). Here we can 'catch in the act' a field-in-the-making, delving into the controversies and demarcations that are almost always at play when a field or discipline is born. Moreover, the central epistemological and methodological concerns in the field of artistic research are likewise central to the study of science and technology – how knowledge and understanding are generated in the interplay between artefacts and human agents and what tools are mobilized to that end.

Since the advance of the social study of science in the 1960s we know that academics in retrospect often conceal the internal contradictions and controversies involving the 'first principles' of a research field and the (usually messy) ways in which the field gained its coherence, secures its stability and sustains its durability. Those contradictions and controversies are partially black-boxed by the field as soon as it is considered to be established. The artistic research field has not yet reached such a closure. In fact, due to the fluidity of its substance, it may never reach nor aspire to such a condition. This makes it an interesting case for the dialogue between artistic research and STS.

One of the major controversies with regard to artistic research concerns the problem of demarcation. Artistic research is positioned at the borderline between the art world and academia. Its substances – artworks and artistic practices – are meaningful in the art world, while at the same time they embody or enact knowledge and insights that function as commodities in academia. A recurrent question is what conditions and criteria should be met in order for artistic practices and artefacts to count as valid vehicles for academic research. And might not the introduction of artistic research into academia alter or amend our understanding of what academic research is? (Borgdorff, 2012; 2018)

A way to approach the problem of demarcation, without reproducing the dichotomies that it constructs, is to focus on what happens when artworks and artistic practices 'travel' from the art world to academia, from the realm of the aesthetic to the realm of the epistemic. What kinds of translation, transformation or transposition happen here (see Schwab, 2018a)? How can we trace an artwork from its indeterminate place in practice – indeterminate because no final description exists of what the limits of that practice are – to its epistemological articulation and impact in scholarly discourse? And how can we retrace the outcome of artistic research to its home in the art world, thereby adding in an infinite loop to the abundance of meaning an artwork might display and convey. Here the problem of demarcation comes down to the problem of reference. Bruno Latour has studied the problem of reference in his description of the chain of transformations taking place between world and words in his article 'Circulating Reference', where he follows the trail of the Amazonian soil to the scientific publication (Latour 1999, pp. 24–80). The transformations involved in the travelling – in our case from art-making, composition and exhibition via documentation and publication to dissemination and discussion – can be described in terms of gain and loss: a trade-off between the particular, the material and the local on the one hand and the general, the discursive and the communal on the other.

In this chapter I will zoom in on one important stage in that chain of transformations: the material passage from the documentation of artistic practice to the publication of research. In my approach to that passage, I will draw on Latour's work on reference, combining it with insights from the Social Construction of Technology (SCOT) programme on the closure of controversies (Pinch & Bijker, 1984). The purpose of this exercise is to illuminate from an STS perspective and, partly based on my own involvement, the internal dynamics of the artistic research-field-in-the-making and to discuss epistemological concerns that are central to both STS and artistic research. In this exercise I will focus on the controversies about that concrete moment in which the documentation of artworks translates into published research.[1]

The Journal and the Catalogue

In 2010, when the institutional anchoring of artistic research in higher education and research was already some ten to twenty years on its way – in some countries earlier than in others – artists were increasingly feeling the need for an adequate platform for publishing outcomes of this kind of research. This led to the founding of the *Journal for Artistic Research* (*JAR*). It was set up as an open-access, peer-reviewed international journal for the publication, dissemination and discussion of artistic research and its methodologies and outcomes in all art disciplines.[2] *JAR* was designed as an enhanced web-based multimedia periodical, where different media formats – texts, images, videos, sound files – could be combined and displayed to make epistemological claims in ways that fulfil the expectations of artists and where the artist/author is in control of how content is displayed. This meant, more specifically, that the journal should enable authors to deviate from the standard format of journal publications. It would be able to display images, sound and text in non-hierarchical ways and would allow a research narrative to be composed and 'read' along a non-linear path. Existing repositories and publication platforms did not meet those requirements. In order to establish such an online academic journal for artistic research, one had to develop new software in addition to editorial policies and review procedures. As a consequence, an open-access, open-source multimedia repository was also developed alongside *JAR*. This Research Catalogue (RC), as the repository was called, was to serve as the technological backbone of the journal. The RC would be a free-of-charge, publicly accessible platform through which authors/artists could self-publish their work or submit their work to *JAR*.[3]

This is the place where I have to more clearly confess my complicity in the birth of both the journal and the repository. I was happy to act as one of the editors of *JAR* from its inception in 2009/2010. Together with Michael Schwab, founder and editor-in-chief of *JAR*, I led a two-year funded project that conceived and developed the first instantiation of the Research Catalogue. My involvement with the enterprise changed after a while. From 2015 to 2019, I was privileged to chair the board of the Society for Artistic Research (SAR), a legal entity initially set up for the sole purpose of making *JAR* and the RC possible. In this chapter, I will reconstruct, as a kind of participant observer, some stages in the RC's development. I will try to strike a balance between my being a part of the process of establishing *JAR* and the RC (including some of the controversies it entailed) and the scholarly distance to these controversies that an analytical perspective requires.[4]

Since 2010, *JAR* and the RC have unfolded in many ways. *JAR* has adjusted its article submission process, experimented with publishing peer-review reports, incorporated network conversations alongside its peer-reviewed publications and, importantly, adjusted the peer-review guidelines by specifically asking reviewers – both artists and academics – to reflect on the potential of a submission and how it might be improved (Schwab, 2018b). The thinking behind the publication policies of *JAR* – that is, what it means for *JAR* to publish artistic practice as research – can be apprehended by inspecting what *JAR* has published over the years. It can also be traced by studying the informative editorials in all issues.

The Research Catalogue will be the focus in my discussion.[5] It is 'owned' by the Society for Artistic Research, and it has developed over the years into a widely used platform for the archiving, documentation, management, publication and dissemination of artistic research. While still functioning as the technological base of *JAR*, the RC is now also used by several other journals as a multimedia platform to handle and display research.[6] An important asset is the use of the RC as an institutional repository. A growing number of European higher education institutes, most notably art schools and art universities, use the RC as their local research portal, through which research by students and faculty is administered and communicated. Together the journals and institutes constitute the Society for Artistic Research's 'portal partners', a group of dedicated RC administrators that meets on a regular basis, exchanges experiences and discusses future adaptations and extensions of the software platform.

Now the RC is mostly employed outside the portals – by artists who use the free-of-charge platform as an archive, as a private or collaborative workspace, as a web channel to share their work-in-progress or as site to self-publish their work in the ever growing network of other artists using the RC. Over the years, the number of users has grown from about 250 in 2011 (when the beta version of the RC was released) to upwards of 12,000 from all over the world in 2019. In what follows, I will describe how the RC operates in the chain that connects artworks and artistic practices with academic publication and evaluation and how this can be illuminated from the perspective of the SCOT programme.

The Artefact

The RC can be described as an instrument in a series of transformations. On the one extreme, there are the works of art made in the studio or the live action on stage or in public space – material tangible or ephemeral artefacts, live performances et cetera. On the other extreme, there is the peer-reviewed publication that circulates in academia. As noted, Latour has discussed the reference between world and words as a chain of transformations or translations, in each stage of which you gain some and lose some (Latour, 1999, p. 70). You gain stability and the potential for distribution at the cost of the singularity and materiality of the operator. In artistic research this involves the chain of reference between artworks at one extreme and artistic research publications at the other. In this series of transformations, the work of art passes through several stages in which the 'common operator' (ibid., p. 69) is matter at one point and form in the next – from material artefact to archived material, from archived material to presentation (for example: exhibition or performance, offline or online), from presentation to documentation, from documentation to publication and so on, whereby you gain some and lose some at every stage.

The RC can be put to use in different stages of this chain. It can be employed as an archive of digitally preserved media distilled from the actual work; as a platform for the web presentation of digital art; as an instrument for the documentation of work, which is always also an exercise of selection; as a web-based studio space to work, alone or collaboratively, on a project; as a publication platform to make a point; or as a public or semi-public channel to share, distribute or discuss works and insights.

In the context of artistic research, the RC is positioned precisely in the gap between the documentation of the work using texts, images and sound and the publication of the work as research. Something happens here that is crucial: a transposition of the work from the aesthetic realm to the epistemic realm. Or rather, a translation or interpretation of the artistic work as research. In the chain of reference from artwork to publication, several earlier transpositions and translations have already been made. To black-box for the moment the making of the art itself, the artistic work goes through a transformation at the moment when it is exhibited or staged. Numerous choices are made about the what, when and how of the presentation of the work. To curate is to translate. And documenting the work – a next step in the chain – also implies a selection with respect to what one wants to convey, aesthetically or otherwise. What images, videos or descriptions will be used to capture the work, in order to gain a specific understanding of it, and what will be lost in that translation? The transition into the epistemic realm is an explicit next move to inscribe the documented work together with other materials (for example: texts) in academia in order to make a claim, to convey knowledge and understanding, shot through with aesthetic experiences. Michael Schwab (2010) has coined the term 'expositionality' for this move from documentation to publication: 'With the notion of 'exposition', we wish to suggest an operator between art and writing. ... [It] is meant as the re-doubling of practice in order to artistically move from artistic ideas to epistemic claims' (Schwab & Borgdorff, 2014, p. 15; cf. Schwab, 2014a; 2014b). If we want to understand how the RC platform works as an instrument for that operation, we should shift our attention for the moment from the question of reference to the RC as an artefact itself, and look at how the platform is produced and used by people and at what controversies were involved in its development. Those controversies, and the work done to overcome them, might tell us something about what it means to expose art practice as research, and hence also about how reference functions in that operation.

The RC can be seen as a 'technological artefact' (Pinch & Bijker, 1984). In the 1980s, the SCOT programme complemented the social study of scientific knowledge and in particular the Empirical Programme of Relativism. The SCOT programme holds that one should look at the social factors and actors involved in technological development – including cultural and political factors and actors – in order to understand why technology works or fails, as well as why technological developments follow mostly non-linear courses in practice. SCOT uses concepts from the Empirical Programme of Relativism as its main analytical tool. There is the principle of symmetry – that is, that any success and any lack of success of technological artefacts should both be treated at the same level and analysed with the same tools and not merely explained after the fact by the supposed superiority or inferiority of the technology. The often erratic course of technological development can be explained by considering the conflicting and constructive roles that social factors and actors play in that development. For analytical purposes, relevant social groups – including the users, the audiences and the producers of the artefact – must be identified, for they play a decisive role in the evolution of the technology. Also, the artefact-in-the-making is not just one thing. Different social groups attach different meanings through the specific ways they use the artefact. The dynamic of technological development is a consequence of this 'interpretative flexibility' (Pinch & Bijker, 1984, p. 409), and it is characterized by controversies about what direction the development should take and what the best solution will be to the problems raised by the various social groups that are involved.

The SCOT programme has meanwhile amended its unidirectional scenario, where the social, cultural and political were constitutive for technological evolution and change. The scenario has been corrected, partly under the influence of actor – network theory, by demonstrating that the principle of symmetry also operates on another level. The social is not the bedrock explanatory force, for the social is itself constituted, staged or assembled in the interplay between – here another symmetry – human and non-human (for example: technological) actors, while the social, in its turn, also constitutes those actors (Latour, 2005). Though keeping that in mind, it still makes sense here to use the early SCOT approach in order to understand how different people have different interpretations and different problems with respect to the RC as a technological artefact and how such diversity co-determines the course of its development. The RC could stabilize over time when 'closure' is achieved – that is, when the controversies over interpretation and use of the artefact itself arrive at a point where consensus exists on a single set of interpretations and uses. But closure does not necessarily have to be reached, as we will see when we discuss the present state and utilization of the RC.

The People and the Problems

So who were the people that came up with the idea of creating a journal for the publication, evaluation and dissemination of artistic research in 2009? And what social groups can be identified that are or have been involved in the development and use of the platform? The two main initiators of the journal, Michael Schwab and Florian Dombois, both worked, as now, in the domain of visual art (photography, post-conceptual art, installations) and had backgrounds in philosophy and science (one in information technology, the other in geophysics).[7] That triple background – philosophy, technology and visual art – was to steer the initial concept and design of *JAR* and the RC platform.

Philosophy, and then specifically theory of knowledge, informed the project from the beginning. It was clear that, related to the problem of demarcation sketched above, epistemological issues come into play when one publishes artistic work as research. What kind of knowledge and understanding does the work embody or convey? And how to present that knowledge and understanding on webpages, using different media in addition to texts? Can one justify non-propositional and non-conceptual forms of knowledge in the context of research? How can such implicit (tacit) understandings, which are generally entwined with aesthetic experiences, be scrutinized in public academic discourse?

Proficiency in areas of technology, in particular information technology and human–computer interaction, was not only helpful in the initial conception of the RC software platform, it also steered the project in a direction that later turned out to have its own difficulties. The technological artefact proved difficult for some social groups to master, and this challenged its usefulness. Musicians, for instance, more often than not lacked affinity with IT instruments, whilst also experiencing the platform as disproportionately attuned to the visual.

In some respects, the visual-arts background of the platform's architects turned out to be decisive for the initial direction of its development. The international consortium that discussed and tested the Artistic Research Catalogue[8] (as the funded project was called at the time) likewise consisted mostly of artists (and curators) affiliated to art schools and art institutes with backgrounds in visual arts – with a few exceptions that would prove relevant at a later stage. Moreover, the development of the RC technology did not start from scratch. The planned software was to be an adaptation of DILPS – Digital *Image* Library Processing System (my italics) – developed as an advanced media archive at the Karlsruhe University of Art and Design.[9]

The three groups involved – initiators, users and developers – were, one could say, somewhat biased towards the visual. Although the intended technological artefact was targeted from the start at displaying images, videos *and* sound alongside text 'on the same plane' on the webpages, the way in which the platform was envisaged betrayed the social (in this case, artistic) backgrounds of the stakeholders. It also revealed the extent to which disciplines matter when it comes to how we understand 'art'. Illustrative is the language initially used to describe the RC webpage by the people involved. It was said to be a 'canvas' on which you can 'draw' your argument, a 'weave' in which material can be 'woven' together, a 'poster' to show your research and to display visually how the different elements relate, or a 'desk' on which material can be 'mapped' and organized (cf. Döbereiner, forthcoming). And in the guidelines for the *JAR* peer reviewers, questions were included about the graphic design of the pages and about whether the visual navigation along the rectangular boxes with texts, images, videos et cetera made sense.[10] The visual nature of the RC arguably facilitated some forms of artistic research exposition more than others.

Soon after the release of the beta version of the platform, the RC was tested in various environments, including the realm of higher arts education, and most extensively in the master's programmes at the Royal Conservatoire in The Hague. Since 2013, the RC had been implemented in the curriculum there to enable music students to 'compose' their master's theses and to publish and disseminate such 'expositions' via the institute's website.[11] This implementation in a music environment incited much feedback on the functioning and dysfunctioning of the platform, as well as suggestions for improvement. The conservatoire in The Hague, for instance, supported the development of a new footnote tool that enabled notes to be inserted as rich-media popovers; the University of Music and Performing Arts in Graz hosted a project that funded an improved media player that promised to deliver a more advanced web audio tool.

Now music students (and teaching staff) are, by and large, rather less visually oriented and also less acquainted with the use of technological devices that fall outside the repertoire of standard consumer electronics, with the notable exception of students and staff in the fields of sound art and sonology. Yet the latter judged the platform from their disciplinary perspective as too limited or too conservative, as did some other artists working in areas of digital or new media art. Students in music are familiar with new media web applications (most have social network accounts) but are hesitant to learn an unfamiliar instrument, let alone to design an argument with that instrument. Such an activity is seemingly far removed from their core field of study, which is usually playing a musical instrument and not a technological one. Although music students and staff have increasingly learned over the years to appreciate the enhanced media possibilities that the RC affords – often by taking notice of successful expositions from earlier cohorts – some of the chief problems with the platform experienced by this social group have persisted and have fed into the controversies to be described below.

Most music students, at the start of their studies, have a rather traditional idea of what research is. They do not initially aim at integrating their own artistic practice into the research design – let alone at making it part of the documented outcome of their research. Instead they tend to fall back on the idea of research as resulting in 'academic writing'. The RC challenges such a text-based conception of research. 'Writing' in the context of artistic research involves the very integration of different media in order to make a claim. That said, this social group felt strongly that the RC was in many respects a hurdle they did not feel the need to tackle. So they often opted instead for uploading traditional text-only documents to the RC, sometimes spiced up with illustrations next to the linear narrative. The reliance on texts goes hand in hand with shying away from the blank page that the RC editor starts out with when an exposition has yet to be created. Such difficulties with the platform translated into the explicit request to be able to work with templates, in which text can be rendered without having to think about the composition of the page or about ways in which content in different media could be displayed. It should be clear that the idea of such 'templates' runs counter to the initial rationale of the platform. Remember that the RC was conceived as a device to display multimedia content in ways whereby the artist/author decides on the layout, the hierarchies and the form and course of the narrative.

But beyond the architects and users portrayed above, an additional social group was involved, and this was to substantially influence the RC's further course of development. These were people representing the interests of higher education institutes. As mentioned above, the RC came to be used more and more as an institutional repository by art schools and art universities in Europe. Research coordinators and local RC portal administrators, along with institutional management representatives who secured funding for adaptations to the RC software, acquired a major influence on the platform's development. Without saying that whoever pays the piper calls the tune, significant amendments to the platform were prompted by the feedback and input of those institutes. For instance, much time and energy (and funding) was put into developing an 'application module' in the RC. It enables funding agencies or art schools to administer and review rich-media applications for project funding or online applications for acceptance to degree programmes, whereby a portfolio can be included in which the research proposal and artistic work are integrated. This allows artist-applicants to communicate their proposals in ways much closer to their actual work. Even so, although the unique assets of the RC could benefit higher education institutions in a variety of ways (education, communication, application procedures), such institutional interests might not always be the primary concern of independent artists who want to expose and share their artistic research.

One of the institutes, not by chance a music academy, expressed the wish to design an alternative workspace, a second editor of the RC that could be used alongside the existing one to cope with problems that users experienced with the platform. In addition to the aforementioned lack of templates to assist students and staff in designing an exposition, there were also other issues with the 'old' editor that might be resolved with the new one. Acknowledging that many users wanted to create expositions that were primarily text-based, with additional other media on the side, the new editor was to take text display as its starting point. The original RC editor indeed had some problems with texts. Different browsers displayed text differently; sometimes scroll bars appeared in one browser, which had carefully been avoided when the exposition was created in another browser. The absolute positioning of media content in rectangular boxes with fixed sizes on the webpage might be good enough for reading from desktops and laptops, but it was not suitable for reading on tablets and smartphones. It was proposed that the new editor should arrange content responsively: that is, the RC would display webpages adequately on different devices and on screens of various sizes. Responsive design would dynamically adjust the positioning of elements on the page to the affordance of each device, including a relative rendering of texts, flexible image size and similar functions – nowadays the common way in which webpages are designed, created and displayed.

Without addressing the pros and cons of responsive design in too much detail, it should be clear that this move from absolute positioning to responsive design constitutes a major change in the way artistic research can be exposed on the RC platform. One of the thoughts behind the old 'graphic' editor was that it matters epistemologically how texts, images, videos and other content are graphically interrelated, although no clear-cut understanding exists about what rules to follow here. But once we abandon this graphic 'logic', what other logics will come in its place? If we lose this asset, what will we gain? Or to phrase the question in the Latourian prose used earlier: how can we rethink and remake the technological artefact in such a way as to avoid breaking the chain of transformations in the unfolding of artistic practice as research? Will a new logic still perform the translation of the aesthetic to the epistemic or might it redefine both limits of the chain, thus affecting our understanding of artistic work or of academic work (or both)?

Here I have to introduce an additional social group that was, and still is, heavily involved in the development of the RC. In 2015, the expansion of the RC endeavour necessitated creation of a RC management structure to support future development. A steering committee was installed, whose members were key users of the RC. One, and later two, Research Catalogue managing officers were also appointed: in consultation with the steering group, they were to provide a road map for future developments and advise the executive board of SAR about where and how investments should be made. Soon after the beginning, observations were made by the RC managing officers, who had a background in sonology and were familiar with computer engineering technologies, that the platform did not adequately align with current developments in information technology and web design. And although the RC software was in principle open-source, the way in which the platform was conceived, constructed and disclosed was not sufficiently inviting to the open-source community. Nor did it have enough appeal to artist-researchers working in fields like media and digital arts or computer music who also possess the technological knowledge, know-how and interest to potentially contribute to the platform's development.

The Closure of a Controversy?

The introduction of the beta version of the responsive text editor in 2018 coincided with a yet ongoing discussion within the core RC community. It addresses the question of how to further transform or redefine the platform so as to align it more closely with the latest developments in information technology – and thereby also to better secure the RC's adaptability, accessibility and sustainability for the longer term. As things look now, the key to that redefinition is located at a more fundamental level. There are suggestions from the community that we should think of the RC exposition as a programmable digital object (Döbereiner, forthcoming), which can be processed in a variety of digital environments (including, but not limited to, graphical ones). It would hence not be based in the first instance on the spatial relations and distribution of content on the page. The logic of such an exposition would be computational not graphical. It would allow the composing of expositions that follow alternative routes and relations, such as sonic ones, whereby the ways and the moments in which media are displayed are staged and performed temporally not visually. The computational logic of the RC exposition as a digital object would also make it easier to flexibly communicate and transact with other environments, such as search engines or other 'libraries', that also work with the abstract relations and ontologies of digital objects. It is unclear, however, how the exposition as a programmable digital object can preserve the quality of being a technological bridge between the documentation of art and the publication of research – unless we begin thinking differently about what 'research' is in the context of the RC.

This is exactly what is being voiced in the current discussion, and here we can capture the epistemological controversy at hand and try to understand the mechanisms that move towards a potential closure – or not – of that controversy. In the debate about the future of the software platform, some people stress the use of the RC as a platform for work-in-progress. The 'open' character of the RC exposition as a digital object empowers its flexibility as a research tool. It invites us, as it were, to see materials as adaptable and portable sketches, as fragmentary notes and scribbles, and to see the exposition as a diary or log book where preliminary thoughts are brought together in an experimental environment where provisional knowledge can emerge. Epistemologically, this points in the direction of a theory of unstable knowledge, or a 'xeno-episteme' (Maharaj, 2004), which incorporates and embraces differences and ambiguities. Methodologically, it points to the RC as a framework for discovery, as an experimental system, which provides the ground for the realization and articulation of indeterminate 'epistemic things' (Rheinberger, 1997). And with regard to the outcome of the research, the mode of its publication and the justification thereof in academic discourse, it points to a suspension or deferral of final results – to an understanding of research outcomes as always work-in-progress, not least where the non-human actors in play, such as browsers, co-determine those outcomes.

The choice to be made in the further development of the RC platform echoes, in a sense, the dichotomy sketched earlier in this chapter, which is also the hallmark of the demarcation problem: does the RC foster knowledge and insights suitable for circulation in academia, or does it serve experimental artistic practice? But perhaps we do not have to choose between the rock of academia and the hard place of the art world. The RC as a bridge in the chain of transformations can surely – precisely thanks to its interpretative flexibility – move from one point to the other in that chain, as long as the reference is not broken (see Figure 2.1). In that respect, there is no stabilization of the RC as a technological artefact. No closure of the controversy is needed.

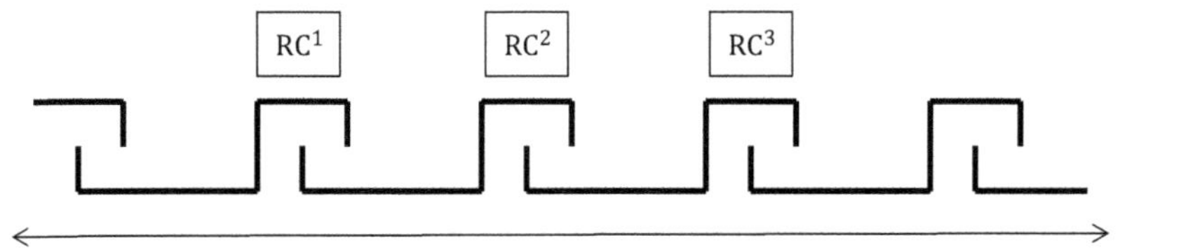

Figure 2.1 An adaptation of Latour, 1999, fig. 2.21. The Research Catalogue as technological artefact that produces transformations in the gaps of the chain of reference between artworks and publications. The numbering (RC^1, RC^2 etc.) does not refer to earlier or later versions, but to different uses of the platform.

Artistic research can be described as a coupling of experimentation and interpretation. The RC nourishes both sides. Just as the contrast between the context of discovery and the context of justification is tempered in contemporary theory of science, particularly in STS, we likewise should not worry about the consistency of the RC throughout the chain of transformations. The RC as technological artefact offers a hybrid and unconventional 'knowing space' (Law, 2014, p. 47) of artistic research, inhabited by various social groups that use the artefact in a variety of ways to experiment with and expose artistic practice as research.

The aim of the exercise in this chapter was to train a spotlight on several stages in the development of the Research Catalogue and to follow the actors involved, in order to say something about what is epistemologically at issue in artistic research. By drawing on Latour's work on reference and using SCOT's toolbox, we were able to unpack the moment in which the artwork passes the gap dividing the aesthetic from the epistemic. At the same time we saw that the controversies about the RC as technological artefact problematized our understanding of where research should be located (cf. Latour, 1998). Do we focus on processes or on outcomes? Should the RC act as an experimental system, facilitating discovery – the emergence of the yet unknown – or should it be geared towards justification of artistic research publications?

The development of the RC, like the field of artistic research itself, has not reached a final closure – and it may never reach, nor even aspire to, such a closure. The work by the RC community on the 'logics' of exposition is itself a research-project-in-progress into the rationale of artistic research. The dispute about the RC exposition that would follow a computational logic – as opposed to an understanding of the exposition as a framework where the positioning of research elements matters to how knowledge and understanding are conveyed – contributes to that wider research project. As such, the artistic research project can inform STS about how a field performs research whilst refusing to place too much confidence in first principles. Such principles would surely reinstall the very demarcations and boundaries that the artistic research project strives to challenge. STS has helped us to understand how the RC, as a hybrid knowing space, negotiates and traverses the boundaries between the art world and academia. At the same time, artistic research exemplifies a central concern of STS – the constitutive role of people, practices and things in the production of knowledge and technologies. It thereby presents to STS an unconventional research-field-in-action.

Notes

1 'Artworks' here denotes all kinds of artistic practices and the artefacts or events that are integral to those practices, such as paintings, sculptures, installations, designs, performances and digital art.
2 www.jar-online.net. For the story of *JAR* and its peer-review policy, see Borgdorff, 2012, ch. 11.
3 The RC website is found at www.researchcatalogue.net.
4 This chapter could not have been written had I had no access to the scribbles, notes, draft papers and discussion contributions of the people involved in the development of SAR, *JAR* and the RC. I am especially obliged to Michael Schwab, Luc Döbereiner, Casper Schipper and Gabriel Paiuk for their thinking about the future of the RC and for their commentary on earlier versions of this text.
5 A more elaborate description of the ins and outs of the RC can be found in Schwab (2014a).
6 *RUUKKU. Studies in Artistic Research*, http://ruukku-journal.fi/en/web/ruukku; *Journal of Sonic Studies (JSS)*, https://www.researchcatalogue.net/view/558606/558607; *VIS. Nordic Journal for Artistic Research*, https://www.en.visjournal.nu.
7 Florian Dombois (https://www.researchcatalogue.net/profile/?person=499); Michael Schwab (https://www.researchcatalogue.net/profile/?person=10953).
8 The project and its consortium are described in Borgdorff, 2012, ch. 11.
9 This initial framework has meanwhile been replaced.
10 See the *JAR* peer-review form at http://www.jar-online.net/peer-reviewing-and-artistic-research.
11 The research portal of the Royal Conservatoire, The Hague, is found at https://www.researchcatalogue.net/view/517228/517229.

References

Borgdorff, H.A. (2012). *The conflict of the faculties: Perspectives on artistic research and academia*. Leiden: Leiden University Press.

Borgdorff, H.A. (2018). Foreword. In J. Wilson, *Artists in the university: Positioning artistic research in higher education* (pp. v–xi). Singapore: Springer.

Döbereiner, L. (forthcoming). The research catalogue exposition as a digital object: Challenges and future. In P. de Assis & L. D'Errico (Eds.), *Artistic research. Charting a field in expansion*. Lanham, MD: Rowman & Littlefield.

Latour, B. (1999). *Pandora's hope: Essays on the reality of science studies*. Cambridge, MA: Harvard University Press.

Latour, B. (2005). *Reassembling the social: An introduction to actor-network-theory*. Oxford: Oxford University Press.

Law, J. (2014). STS as Method. In U. Felt, R. Fouché, C.A. Miller & L. Smith-Doerr (Eds.), *The handbook of science and technology studies* (pp. 31–57). Cambridge, MA: MIT Press.

Maharaj, S. (2004). Unfinishable sketch of 'an object in 4D': Scenes of artistic research. In A.W. Balkema & H. Slager (Eds), *Artistic research* (pp. 39–58). Lier en Boog Series of Philosophy of Art and Art Theory, 18. Amsterdam: Rodopi.

Pinch, T.J., & Bijker, W.E. (1984). The social construction of facts and artefacts: Or how the sociology of science and the sociology of technology might benefit each other. *Social Studies of Science*, *14*(3), 399–441.

Rheinberger, H.J. (1997). *Toward a history of epistemic things: Synthesizing proteins in the test tube*. Stanford, CA: Stanford University Press.

Schwab, M. (2010). Editorial. *JAR 0*. Retrieved from http://www.jar-online.net/issues/0.

Schwab, M. (2014a). Expositions in the Research Catalogue. In M. Schwab & H.A. Borgdorff (Eds.), *The exposition of artistic research: Publishing art in academia* (pp. 92–104). Leiden: Leiden University Press.

Schwab, M. (2014b). The exposition of practice as research as experimental system. In D. Crispin & B. Gilmore (Eds.), *Artistic experimentation in music* (pp. 31–40). Leuven: Leuven University Press.

Schwab, M. (Ed.) (2018a). *Transpositions: Aesthetico-epistemic operators in artistic research.* Leuven: Leuven University Press.

Schwab, M. (2018b). Peer reviewing in the *Journal for Artistic Research.* In W. Ysebaert & B. Van Kerckhoven (Eds.), *Evaluating art and design research: Reflections, evaluation, practices and research presentations* (pp. 52–59). Brussels: VUB Press.

Schwab, M., & Borgdorff, H.A. (Eds.) (2014). *The exposition of artistic research: Publishing art in academia.* Leiden: Leiden University Press.

3 From Quasi-objects to Artistic Components

Science Studies and Artistic Research

Esa Kirkkopelto

The question concerning the relation between science and technology studies (STS) and artistic research (AR) invites a rethinking of the basic principles of inter- and transdisciplinary research (Nowotny et al., 2001; Newell, 2001; Carp, 2001; Barry et al., 2008; Osborne, 2015). It is not a coincidence that both of these research areas focus particularly on procedures they call "translation". In the case of STS, "translation" may refer to gradual processes of problem-solving between human and non-human actors (Callon, 1986; Latour, 1987; Brown, 2002), for instance, or to the hegemonic moment in the organization of actors at which a macro-actor starts to represent the interests of a group (Callon & Latour, 1981, p. 279) or to the phenomeno-technological processes through which scientific research operates with its objects and through which its epistemology develops historically (Rheinberger, 2008, pp. 89–90). Researchers in the domain of AR, on the other hand, discuss ways in which different modes of art and related modes of experience can be mutually translated and how aesthetic or artistic experience can be translated into a knowledge approachable to non-artistic discourses (see e.g. Elo, 2018). In both cases, epistemic processes are understood as ways of potential meaning-making, in other words, as semantic processes that include radically heterogeneous stages and interlocutors. In scientific research this implies that the things under study, which are not necessarily human, are made to speak to us, humans and researchers, so that we can understand them. From the AR perspective, although things and materials, insofar as they are integrated into artistic processes or works, do speak to us, we cannot necessarily understand their talk, let alone translate it into discursive language. The *same* materials or objects are at the outset and simultaneously capable of both of these negative dispositions: muteness and incomprehensibility. However, whereas the former implies a lack of information, the latter connotes information abundance or ambiguity. My aim in this article is to reduce the question of the relation between STS and AR to the level of this fundamental dilemma, which characterizes the relationship between humans and objects. I pose the questions of how, according to STS, objects are constituted in science and how they are constituted in the arts, and furthermore, how these processes and their results resemble each other, are distinguishable from each other and eventually how they may complement each other.

My argument proceeds in two stages. My discussion in the first one is with authors in STS, notably Bruno Latour, not only because he is a prominent figure in that field but also because the arts constitute a constant point of reference in his writing. The comparison between objects of STS and objects of art-making raises the question of twofold agency, which I discuss in the second stage. My point of reference in this discussion is a reading of the classical discourse of Immanuel Kant concerning the "mathematical sublime" in *The Critique of Judgment* (1790). By focusing on the phenomena "quantum" and "measure", which are central to Kant's discourse, I hope to reach the point at which the accepted understanding of *aesthetic* experience is replaced with another kind of understanding that is more *artistic* and more in line with STS insights.

Extended Agency

As my starting point, I have chosen a quotation from Bruno Latour's article published in 2000, in which he articulates the main principle of STS:

> The "unique adequacy" for which ethnomethodologists have fought so strenuously is a very general principle that strictly forbids using any other thing, for instance, a social function, to explain away the insistence, obstinacy or obduracy of a given site .... We cannot emphasize enough the importance of this feature. If a sociologist abandons the idea of replacing, let us say, the second law of thermodynamics by a social factor this law would be supposed to "express", it means that the same is probably true of all the other objects for which we try to provide an explanation. They too resist being a stand-in, and that is no less true of miracles ... fashion, gender, art, than it is of a rotor engine or of a chemical formula. Such is the contribution of STS to social sciences. (Latour, 2000, pp. 112–113)

If the heterogeneity of research objects is not to be done away with but rather respected, the things that are analysed can be left without "explanation" or interpretation by means of some theory-based meta-discourse. It suffices that these things are observed in accordance with the way they connect to each other as well as to their observers, mediated by the technology and the social factors that bring them forth. The object of research is as explicit as the sum of its factors and their sometimes coherent, sometimes discordant tendencies. The factors under study gain relative autonomy not despite but because of one another. It is significant that "art" is mentioned in the above quotation alongside other cultural, technical and scientific phenomena, not as an exception relative to them.[1] I refer to this idea, which gives the factors in question a more independent explanatory status – in other words they explain one another by their way of acting – as "extended agency". It relates to AR in three interrelated ways.

To begin with, it makes understandable the political claim sustaining the emergence of AR as a new kind of academic and epistemic activity. As artists claim their right to knowledge production, art and art-making cease to be mere objects of research and become new kinds of media in the hands of new kinds of social agents. Research makes them less dependent on public opinion and recognition – in other words, the laws of the art market – and expands their role as critical social, political and pedagogical agents. This extension of agency is based on their increasingly discursive (that is, argumentative and critical) and methodological (innovative and inventive) capacities. At the same time, the focus shifts from the accomplished results, or "works", to the modes of working, such as connecting and collaborating with other agents and negotiating with one's medium, materials or surroundings, as well as to modes of exposition capable of communicating these shifts.

Secondly, it concerns artistic authorship and breaks the traditional hierarchies and work distribution that prevail *among* artists. The possibility of conducting research in and through the arts has been welcomed in particular by artists whose role in the production process has tended to be considered secondary in relation to "art work" to be interpreted and its acclaimed "author". In the field of theatre, for instance, AR has had strong emancipatory significance for artists such as scenographers; lighting, sound and costume designers; actors; dancers; singers; instrumentalists and dramaturgs, who have thereby been able to make manifest and emphasize the specificity and importance of *their* artistic agency.

Thirdly, and most importantly in the context of this article, the way in which artists encounter their reality and negotiate with it in practice is, according to STS scholars, analogous to the way scientists conduct their research. The analogy is based not on their results – be it a work of art or a scientific explanation – but on the *logic* according to which they encounter and deal with their objects. This logic, which purports to encounter objects on their own level and on their own terms, could be referred to here as "agential". The models developed in STS may be of help in that artist-researchers tend to understand and articulate the agential logic of art-making.

Even though these three points make some sense if considered separately, problems arise immediately if one tries to understand their connectedness. What do artist-researchers as a new breed of societal and epistemic agent, artists as members of a system of artistic production, and the artistic phenomena operative within artistic practice and studied in AR have in common? How is it possible to consider each one of them as an equal "agent" or "actor"?

Compositions

If one supposes that STS and AR share extended agency, at least to some extent, the next step is to ask where the main differences between them lie. If the idea is to guarantee that the principle of equality between agents holds good within each area and from one area to another, the differences in question cannot be exclusive but should stem from the objects under study or production, from their capacity to "resist the trail" of which also depends their "reality" (Latour 1988, p. 158). The possibility of translatability between fields should originate from the multiplicity, depth and complexity of the things themselves.

The way in which Latour, especially in his later works, stressed the "compositional" aspects of the results of STS provides a possible point of departure for this kind of non-exclusive comparison between scientific and artistic objects:

> Even though the word "composition" is a bit too long and windy, what is nice is that it underlines that things have to be put together (Latin *componere*) while retaining their heterogeneity. Also, it is connected with composure; it has clear roots in art, painting, music, theater, dance, and thus is associated with choreography and scenography. (Latour, 2010, pp. 473–474)

The affinity Latour detects between STS and the arts is intriguing and helps to focus the problem concerning the objective basis of that resemblance. On the one hand, it is clear that the "compositions" retraced and described in STS are not necessarily artistic, even if they can always be illustrated by artistic means and imagining them may produce aesthetic pleasure. On the other hand, an art exhibition curated and composed in an innovative way and inspired by STS may in itself be conceived of as an art work or performance.[2] If the compositions in question are *also* artistic in some way – which they are here, as Latour seemed to think – that aspect must derive from something other than their compositionality. If they have a common point of contact, it must reside in the special way the *components* in each case connect with each other. My hypothesis is therefore that that affinity is based on the nature of the components and their twofold way of connecting, either *experimentally* (in STS) or *artistically* (in AR). The problem, in other words, resides in that particular bifurcation of which the components should be capable and to which they give place. What does this change of perspective, which shifts the focus from the human activity (of a researcher, an artist or an artist-researcher) to the way the objects acquiesce to human observation and/or manipulation, add to this discussion on extended agency? In order to determine this I will next consider the concept "quasi-object" introduced to STS by Latour and question its ontological status as well as its applicability to AR.

Quasi-objects

Latour introduces the notion of quasi-objects in his essay *We Have Never Been Modern* (Latour, 1993, pp. 51–55).[3] He attacks the philosophical foundations that lay behind the positioning and functioning of sociology in systems of knowledge formation, suggesting that it was high time that sociology broke out of the disciplinary framing, which defines it as a study of social facts and leaves the study of material and technological reality to the natural sciences. This breaking away would have radical consequences, the most significant being the overcoming of the metaphysical opposition between nature and culture. If that fundamental opposition were to be abandoned, the reality would be filled with and populated by new kinds of ontologically and epistemologically ambiguous entities "quasi-objects", which he describes as

> simultaneously real, discursive, and social. They belong to nature, to the collective and to discourse. If one autonomizes discourse by turning nature over to the epistemologists and giving up society to the sociologists, one makes it impossible to stitch these three resources back together. (Latour, 1993, p. 64)

If the modernist metaphysical dogmas based on bipolar oppositions are abandoned, what remains is simply a multitude of quasi-objects, the mutual association, organization and co-existence of which would be the greatest political challenge. Former objects of "nature" would turn into political "things", issues of our multi-layered "concern" (Latour, 2000, p. 120; Latour & Weibel, 2005, pp. 22–26). Correspondingly, the basis of political reasoning shifts from ideologies towards everyday practices as well as scientific research, which have never respected the above-mentioned divides in their way of encountering reality and operating with their objects and therefore "have never been modern".

Actor–network theory (ANT), formulated by Michel Callon and Latour, accounts for the function of quasi-objects and their irreducible double nature. Latour connects the concept of quasi-object to ANT thus in an article from 1996:

> If choosing words for the network-tracing activity has to be done, quasi-objects ... or tokens might be the best candidate so far. ... As a rule a quasi-object should be thought of as a moving actant that transforms those which do the moving because they transform the moving object. ... As a rule, what is doing the moving and what is moved have no specific homogeneous morphism. They can be anthropo-morphic, but also zoo-morphic, phusi-morphic, logo-morphic, techno-morphic, ideo-morphic, that is "(x)-morphic". (Latour, 1996, p. 16)

The ontological status of these objects remains an open question. Basically, it should depend on the objects how they conceive of themselves or want to be conceived of (see e.g. Latour 1993, pp. 127–129; Latour, 1996, p. 10). What complicates the issue further is that the author wishes to retain the semiotic nature of these objects and link it to the corresponding post-structuralist idea of the "autonomy of language" (Hoestaker 2005). In the article quoted above Latour acknowledges the debt of ANT to the legacy of post-structural philosophy. The way he defines "actor" associates it with structuralist instances, such as the "floating signifier" à la Claude Lévi-Strauss: "No, what circulates has to be defined like the circulating object in semiotics of texts – especially scientific texts" (Latour, 1996, p. 14; cf. Lévi-Strauss, 1987). As a result, one witnesses the birth of a "new ontological hybrid" that calls for a hybrid ontology:

> Building on the semiotic turn, A[N]T first brackets out society and nature to consider only meaning-productions; then breaking with the limits of semiotics without losing its tool box, it grants activity to the semiotic actors turning them into a new ontological hybrid, world making entities; by doing such a counter-copernican revolution it builds *a completely empty frame* for describing how any entity builds its world. (Latour, 1996, p. 14; italics added)

This idea is relevant from the AR perspective and the way it conceives of its components. Each entity is granted, in addition to agency, a certain viewpoint of the "world" it "builds". This world is in some way meaningful to its builder, a greater or smaller semantic universe, which also makes translation between worlds understandable and necessary. Sometimes the agent is a human being, but usually it is not. What unites them, nevertheless, is the "empty frame", in other words both the agential and the compositional position, which opens up their respective fields of interaction. The familiarity and difference between AR and STS therefore depend on the constitutive ambivalence of their objects, which in order to be artistic also has to be *aesthetic*, at least to some degree: objects of singular and subjective experience.

In the "object-oriented ontology" (OOO) of Graham Harman, whose thought has been significantly influenced by Latour, quasi-objects have an intrinsic structure and dynamics, a sort of inner life, according to which they yield themselves to interaction or withdraw themselves from it. The resemblance and difference between Latour and Harman become apparent in the following citation, in which Harman first associates

ANT with quasi-objects and then gives them an independent ontological status by referring to their "dark inner natures":

> What we find always and everywhere are simply networks of actors. The actor is not quite an object and not quite a subject; or rather, it can behave like both of these, depending on how we view it. Following Serres, Latour makes use of the term "quasi-object" to refer to the precarious status of entities. On the one hand they are contextualized by the objects with which they are fused; on the other, they have retreated into their own dark inner natures and are never fully measured by the networks in which they are involved at any given moment. (Harman, 2010, p. 80)

Harman's interpretation of quasi-objects is ontologically excessive not only from the Latourian perspective, but also from that of AR.[4] In my opinion, what artist-researchers need to do is to open up new ontological perspectives and opportunities without basing their very existence on them, in which case art could easily assume a relatively subjugated and "illustrative" role. The idea of a prevailing symmetry between actors, which is taken for granted in the present discussion, over and again compromises these kinds of ontological temptations. Insofar as an artist-researcher is a hybrid agent the objects he or she encounters and deals with are also hybrids of a sort. How, then, could one conceive of this equality without ontological crutches?

The hybrid objects AR deals with are sometimes referred to as "boundary objects" in methodological discussions (Borgdorff, 2012, p. 177),[5] in other words objects that change their ontological and epistemological nature depending on the research context and which, therefore, can be used for both tracing and questioning disciplinary borderlines. The objects in AR have also been compared to "epistemic things", the "irreducible vagueness" of which constitutes the point of departure for scientific inquiry especially in experimental contexts (Rheinberger, 1997, pp. 28–37; Borgdorff, 2012, pp. 112–120). These "objects of knowledge", as Karin Knorr Cetina calls them, have an "unfolding, dispersing and signifying" nature, and therefore they are always also "partial" in relation to the human desire to know (Knorr Cetina, 2001, p. 184). In this respect, the objects of scientific practice are at least comparable to the artistic elements and components produced by artists. In line with Henk Borgdorff's discussions with Rheinberger and Knorr Cetina, Paulo de Assis recognizes here a clear point of connection with his artistic research project "Musical Experiment 21" (De Assis, 2103–2018), which, among many other subjects, focuses on musical works and their various material "tokens": sketches, manuscripts, editions, recordings, academic and critical reactions, the choice of instruments and the aesthetic orientation of the performer, as well as his or her historically, biologically and culturally informed body:

> Moreover, making explicit the epistemic complexity of musical works allows us to understand works as made up of a myriad of "boundary objects" …. To make performances using selections of such "boundary objects" is an act that discloses open-ended possibilities for new assemblages. Crucial to these new assemblages – and necessary to enhance their epistemic complexity – is the inclusion of a productive "not-yet-knowing", the creation of room for what is yet unthought and unexpected. (De Assis, 2013, p. 157; cf. Borgdorff, 2012, p. 173)

This is clearly a new way of conceiving of artistic compositions both in theory and in practice. However, the statement also sets a certain epistemological challenge for AR: is there a point at which the "productive not-yet-knowing" finally turns into something "already known" or, in Rheinberger's terms, at which an "epistemic thing" changes into a knowable and controllable "technical object"? Should one not rather think that artistic objects, even at the end of a research project, remain in essence "unfolding, dispersing and signifying", to quote Knorr Cetina (cf. Borgdorff, 2012, pp. 197–198)? This is what de Assis implies as well. In his recent work he discusses the double nature of artistic compositions as "virtual structures and actual things":

> When looking at those exploded things, a musician or a scholar has two options: one is analytical, remaining at a certain distance from the materials of musical practice, questioning things in terms of what they are, how they appear, which properties they have, and what relations they entertain with each other; the other option is one that decidedly dives into the materialities of music-making, focusing on what to do with these things, how to reactivate them, searching for the yet unseen *virtual components* that they possess, asking which potentialities they have, how to give them renewed sounds and furies, and how to express them anew. (De Assis, 2018, p. 41; italics added).

Keeping this conclusion in mind I now return to my starting point concerning the dual way in which objects may reject access to them, also with reference to the above discussion with Latour. Thus far, the components of both regimes, STS and AR, are semantic, albeit semantically open, which implies that at times they may also appear semantically empty (mute or ambiguous). Comparing the fields has produced a seemingly modest but significant shift. The relation between them is no longer based on the metaphysical and "modern" opposition between the empirical and the aesthetic, their "primary" and "secondary qualities" as Hume or Locke might call them: it is based on an affinity between the experimental and the artistic, which is not the same thing. The latter difference refers to the way objects are encountered and articulated in two different kinds of practice, scientific and artistic, and this difference now is real. For the same reason, instead of ignoring it or consolidating it by establishing metaphysical divides between different kinds of epistemic or ontological domains, I will argue in favour of the idea that the objects themselves, insofar as they can be considered Latourian actors *and* artistic components, appear originally in a double register.

Measures

My aim is now to show how experience comprises actual moments, when the empirical and the aesthetic are mixed in an indecisive and therefore constitutive manner. Giving primacy to these moments makes it possible finally to find a perspective from which the twofold appearance of the components, as both experimental and artistic, becomes conceivable. The moments in question relate to measurement and, in particular, to the units and their empirical counterparts, the quanta.

Let me start with a quotation from the German poet and philosopher Friedrich Hölderlin, someone who, today, could be considered an artist-researcher *avant la lettre*. In his posthumously published poem, written at the beginning of the nineteenth century and later entitled "In lieblicher Bläue" ("In Lovely Blue"), Hölderlin addresses the question of measurement in the following ways:

> Ist unbekannt Gott? Ist er offenbar wie die Himmel?
> dieses glaub' ich eher. Des Menschen Maaß ist's. (Hölderlin, 2004, p. 23)
>
> Is God unknown? Is he manifest as the sky?
> This I tend to believe. Such is man's measure. (Hölderlin, 1984, p. 249)

And a little later in the same poem:

> Gibt es auf Erden ein Mass? Es gibt keines. (Hölderlin, 2004, p. 23)
>
> Is there a measure on Earth? There is none. (Hölderlin, 1984, p. 250)

The poet's conclusions are particularly interesting given their historical context: the aftermath of the Kantian Copernican revolution and the connection the second one establishes between human experience and the physical laws governing heavenly mechanics. Against this background, the loss of measure he refers to could be understood in both scientific and poetic terms. In that there is no (given) measure, we humans have to measure *both* in science *and* in the arts.[6] Measurement is conceived of as a human activity distinct from the universe it describes and not as deriving from some stable and given metaphysical principles or models (Perhoniemi, 2014). It literally has a fundamental significance and function as our only reliable way of relating to something that exists independently of our judgement and, in many cases, our occasional existence. According to Hölderlin, human existence, its success or ruin, as his contemporary readings of the tragedies of Sophocles indicate, is dependent on the capacity to measure (Kirkkopelto, 2012).

Here is a more up-to-date prosaic definition of the same subject: "Measurement – a central epistemic activity in science – relates a number and a quantity in an effort to estimate the magnitude of that quantity" (Trout, 2000, p. 265). This is an ambiguous contemporary formulation, which dates back to Aristotle and links concepts such as "number", "quantity" and "magnitude", as if one already knows what each one of these terms means. Allow me to leave their meanings in the air for a while and consider the issue from a perspective that links it to the previous discussion. I will confine my considerations to Kant, for whom the question of the transcendental conditions of measuring clearly constituted a boundary case between empirical and aesthetic judgement.

Quanta

Kant deals with the topic on two occasions in his *Critiques*. In the first *Critique* (Kant, 1998), he analyses the way in which empirical reality is born and appears to us as something a priori quantifiable (see "The Axioms of Intuition", ibid., pp. 286–289). Secondly, he returns to the issue in the third *Critique* (Kant, 1987), in the chapter entitled the "Analytic of the Sublime", where he discusses the type of sublime he significantly calls "mathematic". The two analyses are interrelated, and the latter one also contains explicit references to the former.

The mathematical sublime covers the special area of aesthetic reflexion in which the impressive effect of a phenomenon is based on its magnitude. This is thus different from empirical judgement, meaning that a phenomenon to be measured may continue limitlessly and regardless of the scale of human perception. It is crucial in the sublime experience that the phenomenon to be observed is not too big or too small, too distant or too close, such that it can *challenge* the above-mentioned scale. This border-line or sub-liminal situation brings human cognition into contact with the *claim of measure* within it, which Kant identifies with the power of Reason. In the mathematical sublime, the imagination *fails* to represent the magnitude of the phenomenon and thereby indirectly communicates the power that requires the result. The aesthetically and perceptually unsatisfied claim is compensated by the unimaginable idea of "absolutely large", or infinity as a whole. At the same time as the observation of relative *quantity* changes into an aesthetic idea of absolute *quality*, the empirical process of measuring confronts its aesthetic parameters. This is expressed explicitly in the first passage of §26, which I quote in its entirety below:

> Estimation of magnitude by means of numerical concepts (or their signs in algebra) is mathematical; estimation of magnitudes in mere intuition (by the eye) is aesthetic. It is true that to get determinate concepts of *how large* something is we must use numbers (or, at any rate approximations [expressed] by numerical series progressing to infinity), whose unity is [the unit we use as] the measure; and to that extent all the logical estimation of magnitude is mathematical. Yet the magnitude of the measure [*die Große des Maßes*] must be assumed to be known. Therefore if we had to estimate this magnitude also mathematically, i.e., only by numbers, whose unity would have to be a different measure, then we would never have a first or basic measure, and hence also could have no determinate concept of the given magnitude. Hence our estimation of the magnitude of the basic measure must consist merely in our being able to take it in [*fassen*] directly in one intuition and to use it, by means of imagination, for exhibiting numerical concepts. In other words, all estimation of the magnitude of objects of nature is aesthetic (i.e. determined subjectively rather than objectively). (Kant 1987, §26, p. 107)

Although Kant asserts that measurement, as an empirical estimation of magnitude, has an intrinsic aesthetic aspect insofar as it is forced to use measures, in other words units that human perception can "grasp directly in one intuition", this does not make the operation aesthetic through and through. The aesthetic unit of measure in the first *Critique*, insofar as it serves the purpose of theoretical understanding, is provided by the transcendental function of imagination, the so-called "schematism". The imaginary "schema" that unites the theoretical concept (or "category") of *quantity* with the corresponding intuitions of time and space is "number" (Kant, 1998, p. 274). Given the divided nature of transcendental imagination, which sometimes works in the service of theoretical understanding and sometimes appeals aesthetically to Reason, the two kinds of judgement, theoretical and aesthetic, should never be confused. This, at least, is what Kant seems to propose. However, a closer reading of the descriptions of the "Analytic of the Sublime" complicates this divide in an interesting way. As Kant defines it, magnitude is an intrinsic aspect of any perceptible phenomenon insofar as it constitutes a unity, a "thing": "That something is a magnitude (*quantum*) can be cognized from the thing itself without any comparison of it with others, namely, if a multiplicity of the homogenous together constitutes a unity" (Kant, 1987, p. 103).

Magnitude is intrinsically linked to the capacity of a phenomenon to present itself as an entity with a certain unity, which does not yet make it a conceptually determined empirical object. The instance or faculty in charge of fashioning sensorial influence into perceptible data is, as mentioned above, transcendental imagination. Its functioning in both *Critiques* is understood as an interplay between two concomitant processes, "apprehension" (*Auffassung*, *Apprehensio* [Kant, 1994, pp. 239, 275, 296; Kant,1987, §26, p. 108, §27, p. 116]) and "comprehension" (*Zusammensetzung*, *comprehensio aesthetica* [Kant 1987, §26, p. 108, § 27, p. 116]). *Apprehension* packs the cognized sensory material, the perceptions, into an unending temporal and associative series of data that *comprehension*, in turn, recollects and gathers into recognizable phenomena, in other words, unities susceptible to conceptual determination (such as empirical objects) *or* aesthetic reflection. This imaginary elaboration of the sensible starts on the level of the pure intuition of time, insofar as it constitutes our "internal sense" and consists of the "auto-affective" production of instants following each other in an irreversible order (Kant, 1994, pp. 189–190). No sequence of time or spatial perception reaches human understanding without this kind of fundamental imaginary treatment. Observe how, according to Kant, imagination operates in the case of measurement:

> In order for the imagination to take in a quantum intuitively, so that we can then use it as a measure or unity in estimating magnitude by numbers, the imagination must perform two acts: apprehension (*apprehensio*), and comprehension (*comprehensio aesthetica*). Apprehension involves no problem, for it may progress to infinity. But comprehension becomes more and more difficult the farther apprehension progresses, and it soon reaches its maximum, namely, the aesthetically largest basic measure for an estimation of magnitude. (Kant, 1987, §26, 108)

The sublime experience is therefore attributable to a certain crisis of comprehension. Kant occasionally (Kant, 1987, p. 111) attempts to distinguish between "logical" comprehension (*comprehensio logica*) in the service of understanding and mathematical measuring and "aesthetic" comprehension (*comprehensio aesthetica*) in the case of aesthetic judgement, as if there were two different ways in which imagination can function. However, as I see it, a few features of Kantian analysis tend to blur this delicate and abstract distinction based on an attempt to *imagine* how imagination *actually* works. To begin with, apprehension and comprehension always seem to collaborate. Apprehension as a sensible and associative synthesis is already some sort of comprehension, in relation to which comprehension per se is like "squared" apprehension.[7] This implies that the process, on which our perception of time and space is based, is both continuous and discontinuous, in other words a real process of *production*, alternating between generation and recollection. Secondly, qualitative or aesthetic comprehension is at work in the case of theoretical understanding, charged with giving the experience a *qualitatively* coherent and integral appearance, a unity that is called the "unity of apperception" in the first *Critique* (Kant, 1994, pp. 248–251). This transcendental function, which presents the experience to itself as a totality, constitutes the necessary qualitative condition of every theoretical judgment. The empirical and conceptual definition of the sensible, the so-called "a priori synthesis", is only conceivable within this kind of aesthetic dialectic between the unity and the whole, which by definition goes beyond conceptual determination.

If quanta emerge as a result of these transcendental and imaginary operations it does not matter whether they are submitted later to numerical determination or if they are reflected aesthetically, because they basically take shape *before* they encounter human understanding, its empirical concepts or aesthetic judgements. In both cases the phenomena are supposed to be *both* continuous *and* divisible according to their spatio-temporal size, and this constitutes the basis of all their further representations and determinations. Insofar as the sensible data arrives at human cognition in the form of *quanta* and, therefore, as something "given" (*datum*), one could conclude that the servile and spontaneous aspects of imagination mix in the quantum, and that the functional division between the different modes only appears later. Such experience, according to Kant, is *inherently quantified.*

What does Reason finally add to this dynamic? The rational "requirement" of quantum is absolute, in assuring us that whatever we encounter in the universe is just another quantum. How can one be so sure about this? How does one know it a priori? The answer, it seems to me, is simple: because we, their "observers", are some sort of quanta as well. The self-division of the transcendental imagination, which according to Kant takes place on the occasion of both empirical and aesthetic experience and always before being conscious of it, is a process that traces or draws the entities apart and communicates them to each other. In terms of the sublime experience, therefore, in the end it is irrelevant whether a phenomenon is minuscule or gigantic. What communicates itself sublimely is the "power of measure", or the "measuring", in other words the fundamental process according to which things articulate themselves as separable, countable and articulable entities and manifest themselves to each other. If there is still something absolute, it is not the size, or even its requirement, it is the order or the medium (Reason, *logos*) whereby things originate as quanta. This order, which makes things equal enough to be measured against one another, reaches a priori beyond human cognition and therefore also facilitates its extension.

What indirectly betrays the absolute nature of quanta, in other words their *ultimate* givenness, is the way Kant (1987, §25) refers to "measure" at the same time as the "magnitude of a unity" (*Größe de Einheit* [ibid., p. 104]) and as the "unity of a magnitude" (*Einheit de Größe* [ibid., §26, p. 110]). This is perhaps neither a coincidence nor a careless mistake. It only indicates that there is no unity without magnitude and vice-versa. This is how things are given or how they give themselves to us and to each other before or beyond the divide between the aesthetic and the empirical, in their simultaneous muteness and ambiguity. What one should think of such conclusion? If we do not have any imaginable, not even *mathematical*, means[8] for deciding whether this ambiguity and muteness belong to the entities "themselves" or whether they only appear "for us", in other words if we as human observers add them, then we have to take them as an intrinsic aspect of reality that is common to us as humans and to our non-human companions.

Conclusions

Given that measurement is one of the basic operations of science, it is clear that these operations also rest on the supposition of a basic kind of givenness concerning how the entities under study, the "epistemic things", are susceptible to be translated into research data. However, if the standpoint of STS in its Latourian version is valid, this translatability cannot be merely our own doing. It is not necessarily the doing of the entities either. It is rather something that happens between both and according to which we appear as quasi-objects *to each other* and communicate or collide with each other. Observation changes us as observing subjects as much as it changes the objects observed. How should the logic of this transformation be conceived? On the basis of all the above, one could outline the following quasi- or para-ontological model.

An object is *virtual*, something "quasi", to the degree that it is networked. It is *actual* insofar as its trajectory is traceable (cf. Kirkkopelto, 2016). However, an actor always necessarily and simultaneously belongs to more than one series of actors. This is the condition of its relative freedom in relation to other series. If the actor was entirely "here" or "present" in one series, he, she or it would not be free to act in any of them. As Latour recurrently emphasizes, there is nothing outside a network – except, I would like to add, other networks, which in relation to the one observed remain virtual. No external observer can perceive all the connections an actor may have, nor can the number of series to which he or she belongs be defined. Even the actors themselves cannot do that, in that they tend to belong to series unknowingly or unconsciously. Similarly, insofar as actors cannot give themselves entirely *to themselves or to others*, their existence is quasi-autonomous. As a *component* they can only exist. Their existence is constitutively partly concealed partly manifest, partly virtual, and partly actual, not because of themselves – as in Harman's ontology – but because of their connectedness.

Correspondingly, and here again corresponding to the conclusions of De Assis, one could formulate the relation between STS and AR in the following way: STS research focuses on actual compositions, whereas AR is more concerned with virtual aspects – which, as Gilles Deleuze points out (1998), are no less real. Instead of constituting actor-networks, always linked to actual interaction, the objects appear artistically from their virtual and therefore *componential* angle. What distinguishes an artistic from a mere aesthetic object is not, as Kant and the phenomenologists following him suggest, human intention, it is the manner of its connectedness. An *artistic* object is always compositional and componential, whereas an *aesthetic* object is isolated and cut off from its series and presented to a reflective subject as something as mysterious as the subject itself and, thereby, as its pledge. The divide between "empirical" and "aesthetic" reflects the modern divide that the idea of extended agency developed here deconstructs.

The above comparison reveals the limitations of Kantian aesthetics as well as its ulterior idealistic or phenomenological versions, at least insofar as they all lean on the supposition of the "receptiveness" of human experience in relation to what it encounters, regardless of whether it consists of sensory data or works of art. But artistic perception is not like aesthetic reflection. Contrary to what Kantian metrology implies, the human mind is not a "detector", whose objects consist at the first stage of some sort of "matter" to be articulated and made meaningful. Instead, as I have argued, the world and its inhabitants access the realm of (human) consciousness as different kinds of relatively articulated entities, as quanta, as bodies with both extensive and intensive qualities, which enter into associative play with the new entity they encounter. What appears first and last, at the limits of human cognition, are virtual objects that become more actual the more the entities encountered become acquainted with and informed by each other. Like subatomic particles – the analogy is unavoidable here[9] – they appear simultaneously as continuous waves and discontinuous bodies. It is precisely this ambiguity that makes them capable of being experienced, in other words *touched*, by each other and by themselves. It is not the case that first there are entities who then communicate with each other by emitting quantified information: these entities are already quantified, otherwise what they emit could never work as their reliable measure. Or, as Aristotle noted, the unit of measure is always "akin" (*syggenes*) to the thing to be measured (Aristotle, 1924, 1053a 25).

Hierarchical divisions contested by STS, such as between micro and macro, close and distant, have never really existed in the arts, either. The mathematical sublime, as well as its "dynamic" variation (cf. Kirkkopelto, 2014), holds only in modern Kantian metrology, in which the power of imagination, even in its artistic use, is a priori subordinated to the service of comprehending Reason, the human mind encountering the sensible "nature" that awaits its human appropriation and purpose (cf. Latour, 2011). These types of experience deserve reconsideration in the era of planetary environmental crisis: they should either be abandoned, or deconstructed and understood in a radically new way, as demonstrating the processes of apparition that no longer respect the divisions between human and non-human and the related aesthetic frames or scales.

Notes

1 This seems to be a constant feature in Latour's thinking. Even in 2013, art works are displayed alongside other cultural objects as examples of "objects of fiction" (Latour, 2013, pp. 233–257).

2 The most famous examples of this are probably the "Making Things Public" exhibition curated by Bruno Latour and Peter Weibel in 2005 (Latour & Weibel, 2005), as well as the interdisciplinary research program SPEAP (Latour, 2011). Both of these could be observed from the AR perspective, even though they are not defined as such.

3 As Latour notes, the concept derives initially from Michel Serres (1982).

4 In order to reach that level of ontological certainty Harman, unlike Latour, abandons the "linguistic turn" initiated by the (post-)structuralists (Cox et al., 2015, p. 20) This, as I have argued elsewhere, impedes the applicability of OOO in artistic practice, which cannot do without a deconstructive idea of language, capable of functioning both discursively *and* poetically. See Kirkkopelto, 2016.

5 According to Borgdorff, his idea of "boundary object" is inspired by Thomas F. Gieryn's "boundary work" (1983). The same term has also been used by Star and Grisemer (1989) to refer to necessary technical and material devices and tools for sustaining and advancing research practice.

6 In that Hölderlin refers to measure (*Maas*), there is no doubt about his awareness of the results of contemporary science. He was probably also aware of the principles of Newtonian mechanics, not only via Kant, of whose work he was a keen reader, but also more directly. As Wolfgan Schadewaldt argues, the curious expression "eccentric path" the poet uses to depict the dramaturgy of his poetic figures derives from the study of the trajectories of comets (Schadewaldt, 1952). On the concept of measure in Hölderlin, see Polledri, 2002.

7 In the first *Critique* both functions are dealt with under the heading of "apprehension" (cf. Kant 1994, p. 261), even though the associative and recollecting phases are clearly distinguished ("synthesis of apprehension" and "synthesis of reproduction" [Kant 1994, pp. 228–230]) especially in the 1781 version of the work. One might well ponder on how the omission of this distinction in the 1786 version paves the way for its reappearance in the "Analytic of the Sublime". For a further analysis of this point, see Kirkkopelto, 2008, pp. 350–365.

8 At this point I am reminded of Quentin Meillassoux's (2006) argument, according to which only mathematics and mathematically formalizable sciences can reach beyond the transcendental closure of human experience.

9 Karen Barad (2007) makes extensive ontological deductions based on discoveries in the area of quantum physics. Here, and for the same reasons as Harman gives (see n. 4 above), I would maintain distance from them.

References

Aristotle (1924). *Metaphysics I–II*, ed. W. D. Ross. Oxford: Clarendon Press.

Barad, K. (2007). *Meeting the Universe Halfway: Quantum Physics and the Entanglement of Matter and Meaning.* Durham, NC, and London: Duke University Press.

Barry, A., Born, G., & Weszkalnys, G. (2008). Logics of Interdisciplinarity. *Economy and Society, 37*(1), 20–49.

Borgdorff, H. (2012). *The Conflict of the Faculties: Perspectives on Artistic Research and Academia.* Leiden: Leiden University Press.

Brown, S. D. (2002). Michel Serres: Science, Translation and the Logic of the Parasite. *Theory, Culture and Society, 19*(3), 1–27.

Callon, M. (1986). Some Elements of a Sociology of Translation: Domestication of the Scallops and the Fishermen of St Brieuc Bay. In J. Law (Ed.), *Power, Action and Belief: A New Sociology of Knowledge?* (pp. 196–223). London: Routledge.

Callon, M., & Latour, B. (1981). Unscrewing the Big Leviathan: How Actors Macro-structure Reality and How Sociologists Help Them to Do So. In K. Knorr Cetina & A. Cicourel (Eds.), *Advances in Social Theory and Methodology: Towards an Integration of Micro- and Macro-sociologies* (pp. 277–303). Boston, MA: Routledge.

Carp, R. M. (2001). Integrative Praxes: Learning from Multiple Knowledge Formations. *Issues in Integrative Studies, 19*(2001), 71–121.

Cox, C., Jaskey, J., & Malik, S. (Eds.) (2015). *Realism Materialism Art.* Berlin: Sternberg Press.

De Assis, P. (2013). Epistemic Complexity and Experimental Systems in Music Performance. In M. Schwab (Ed.), *Experimental Systems: Future Knowledge in Artistic Research* (pp. 151–163). Leuven: Leuven University Press.

De Assis, P. (2013–2018). Music Experiment 21: Experimentation versus Interpretation. Exploring New Paths in Music Performance in the Twenty-First Century. Retrieved from https://musicexperiment21.eu (accessed 10 July 2019).

De Assis, P. (Ed.) (2018). *Virtual Works – Actual Things: Essays in Music Ontology.* Leuven: Leuven University Press.

Deleuze, G. (1998). *Difference and Repetition* [1968], trans. P. Patton. New York: Columbia University Press.

Elo, M. (2018). Ineffable Dispositions. In M. Schwab (Ed.) (2018). *Transpositions: Aesthetico-Epistemic Operators in Artistic Research* (pp. 281–296). Leuven: Leuven University Press.

Gieryn, T. E. (1983). Boundary Work and the Demarcation of Science from Non-Science. *American Sociological Review, 48*(6), 781–795.

Harman, G. (2010). *Towards Speculative Realism. Essays and Lectures.* Winchester, Hants., and Washington DC: Zero Books.

Hoestaker, R. (2005). Latour: Semiotics and Science Studies. *Science and Technology Studies, 18*(2), 5–25.

Hölderlin, F. (1984). *Hymns and Fragments*, trans. R. Sieburth. Princeton, NJ: Princeton University Press.

Hölderlin, F. (2004). *Sämtliche Werke: Briefe und Dokumente in zeitlicher Folge*, ed. D. E. Sattler, vol. 12 (pp. 22–24). Munich: Luchterhand.

Kant, I. (1987). *Critique of Judgment* [1790], trans. W. S. Pluhar. Indianapolis: Hackett Publishing Company.

Kant, I. (1998). *Critique of Pure Reason* [1781/1786], trans. A. W. Guyer & P. Wood. Cambridge: Cambridge University Press.

Kirkkopelto, E. (2008). *Le théâtre de l'expérience: Contributions à la théorie de la scène.* Paris: Presses de l'Université Paris-Sorbonne.

Kirkkopelto, E. (2012). Hölderlin, Sophocle et les deux rythmes de la modernité, *Poésie*, 141, 100–113; English trans.: Hölderlin, Sophocles and the Two Rhythms of Modernity. Retrieved from https://www.academia.edu/36295658/Hölderlin_Sophocles_and_the_Two_Rhythms_of_Modernity (accessed 10 July 2019).

Kirkkopelto, E. (2014). Farewell to the Sublime? Performance Criticism in the Age of Terrorism. *Forum Modernes Theater*, 29, number 1–2, Special Issue: "Theatre as Critique." Nikolaus Müller-Schöll & Gerhard Siegmund (eds), 47–55.

Kirkkopelto, E. (2016). Joints and Strings: Body and Object in Performance. *Performance Philosophy Journal*, 2(1), 49–59.

Knorr Cetina, K. (2001). Objectual Practice. In T. R. Schatzki, K. Knorr Cetina & E. von Savigny (Eds.), *The Practice Turn in Contemporary Theory* (pp. 175–188). London: Routledge.

Latour, B. (1987). *Science in Action*. Cambridge, MA: Harvard University Press.

Latour, B. (1988). *The Pasteurization of France*. Cambridge, MA, and London: Harvard University Press.

Latour, B. (1993). *We Have Never Been Modern* [1991]. Cambridge, MA: Harvard University Press.

Latour, B. (1996). On Actor-Network Theory. A Few Clarifications Plus More Than a Few Complications. *Soziale Welt, 47*, 369–381. Retrieved from http://www.bruno-latour.fr/sites/default/files/P-67%20ACTOR-NETWORK.pdf (accessed 9 Feb. 2019).

Latour, B. (2000). When Things Strike Back: A Possible Contribution of "Science Studies" to the Social Sciences. *British Journal of Sociology, 51*(1), 107–123.

Latour, B. (2010). An Attempt at a 'Compositionist Manifesto'. *New Literary History, 41*, 471–490.

Latour, B. (2011). Waiting for Gaia: Composing the Common World Through Arts and Politics: A Lecture at the French Institute, London, for the launching of SPEAP (the Sciences Po Program in Arts and Politics). Retrieved from http://www.bruno-latour.fr/sites/default/files/124-GAIA-LONDON-SPEAP_0.pdf (accessed 8 Feb 2019).

Latour, B. (2013). *An Inquiry into the Modes of Existence. An Anthropology of the Moderns*. Cambridge, MA: Harvard University Press.

Latour, B., & Weibel, P. (Eds.) (2005). *Making Things Public: Atmospheres of Democracy*. Cambridge, MA: MIT Press.

Lévi-Strauss, C. (1987). *Introduction to the Work of Marcel Mauss* [1950]. London: Routledge & Kegan.

Meillassoux, Q. (2006). *Après la finitude: Essai sur la nécessité de la contingence*. Paris: Seuil.

Newell, W. H. (2001). A Theory of Interdisciplinary Studies. *Issues in Integrative Studies, 19*, 1–25.

Nowotny, H., Scott, P., & Gibbons, M. (2001). *Re-thinking Science: Knowledge and the Public in an Age of Uncertainty*. Cambridge: Polity Press.

Osborne, P. (2015). Problematizing Disciplinarity, Transdisciplinary Problematics. *Theory, Culture and Society, 32*(5–6), 3–35.

Perhoniemi, T. (2014). *Mitan muunnelmat: Miten määritämme maailmaa, ihmistä ja tietoa* [Variations of the Measure: How We Determine the World, Human Beings and Knowledge]. Tampere: Vastapaino.

Polledri, E. (2002). *"...immmer bestehet ein Maas": Begriff des Masses in Hölderlins Werk*. Würtzburg: Königshausen & Neuman.

Rheinberger, H.-J. (1997). *Towards a History of Epistemic Things: Synthetizing Proteins in the Test Tube*. Stanford, CA: Stanford University Press.

Rheinberger, H.-J. (2008). Translating Derrida. *Dalhousie French Studies, 82*, 85–91.

Schadewaldt, W. (1952). Das Bild der exzentrischen Bahn bei Hölderlin, *Hölderlin-Jahrbuch* 6, 1–16.

Serres, M. (1982). *The Parasite*, trans. L. Schehr. Baltimore, MA: Johns Hopkins University Press.

Star, S., & Griesemer, J. (1989). Institutional Ecology, "Translations" and Boundary Objects: Amateurs and Professionals in Berkeley's Museum of Vertebrate Zoology, 1907–39. *Social Studies of Science, 19*(3), 387–420.

Trout, J. D. (2000). Measurement. In W. Newton-Smith (Ed.), *A Companion to the Philosophy of Science* (pp. 265–276). Oxford: Blackwell.

4 A Thought Experiment on Artistic Research as High-Risk Ethnography[1]

Ruth Benschop

Introduction

Even though the term has been around for a while, "artistic research" remains a combustible combination of worlds that are often at first thought to be far apart: art and academia. Even though a familiar concept to some and a practical reality in terms of offered courses, institutional underpinning, journals for publication et cetera, to others, it remains a notion that instantly triggers fierce questions: practical, institutional and philosophical. Important questions, to be sure, but nonetheless questions that I will leave aside here. Instead of stepping into a debate about artistic research, and carving out a position about it, in this chapter, I want to sidestep such foundational positioning in favor of conducting a thought experiment that allows us to focus on practice.

This move of sidestepping foundational debates and turning towards practice is typical of the interdisciplinary field of science and technology studies (STS). In STS researchers do not so much conceive of the philosophical and political questions concerning the growing scientific and technological character of society as a search for guidelines or protocols but as an invitation to do empirical, anthropological research of practices in order to arrive at thick (normative) analyses and questions. STS thus developed into a way of postponing as well as feeding prescriptive and normative arguments about science and its societal role. Instead of taking the role of armchair philosopher or arbiter, STS research paid attention to the multiple, heterogeneous and often very mundane ways in which science was actually done, facts became true and knowledge applicable. And it showed the various ways in which science-in-the-making (Latour, 1987) was necessarily and habitually intertwined with other domains and concerns. The objects of STS studies gained specific complexity, materiality, contexts, biographies. This "denaturalization" of science at once placed scientific practices and the knowledge they produce on a par with other practices, as well as making space for alternative realities to be imagined. It thus shifted the ways in which normative questions about science could be addressed, what such normative debates ought to be about as well as the language in which we are able to speak about science.[2] It is this shift that is important for artistic research today.

The necessity of this turn to practices of artistic research, becomes apparent when we look at struggles to describe, critique or support artistic research. Such debates can be understood as attempts to establish and legitimize the research paradigm or discipline for artistic research. (For examples, somewhat arbitrarily, of debates about and attempts to define artistic research, see, for instance, Biggs & Karlsson,

2010; Hannula, Suoranta & Vadén, 2005; Klein, 2012; Nelson, 2013; Slager, 2004; Wilson & van Ruiten, 2013; for analyses of such debates, see, for instance, Benschop, 2014; Spronck, 2016). I want to focus here briefly on an underlying dichotomy that reoccurs in them. A problematic dichotomy described also by Borgdorff (2012) when he discusses critiques of artistic research:

> The problem with this type of criticism is that it fabricates its own object of criticism. It begins by constructing a caricature of artistic research in academia – it is disciplining, homogenising, restrictive, conformist, naive. After that, it is no longer difficult to field a whole line-up of post-Nietzschean witnesses to lambast those pernicious practices, which are inimical to art and which, under pressure from an equally maleficent education policy, are seen to have infected the art world under the label "academisation" in order to subject art practices to their disciplining forces. Such argumentation often follows the same pattern: first you create an antithesis between (inadequate) academic research and the liberating cognitive practices of artists, and then you go on to defend the latter from unwarranted institutionalisation and normalisation. This pattern is the mirror image of another sort of reasoning that likewise posits an antithesis between artistic research and academic research. (Borgdorff, 2012, p. 5)

Regardless of whether authors take a radical position or proclaim artistic research as an intermediate practice, in attempts to define artistic research several dualities reappear: the differences for instance between the arts and academia, between arts education and the university, between theory and practice, between art's autonomy and science's methodology. It is in relation to these that artistic research is then described and positioned. These are foundational skirmishes and border conflicts about the notion that artists do research. To proponents of artistic research, it is of great importance to stress the absence of scientific methods in artistic research. At the same time, critics of artistic research also evoke the divide between artistic freedom and methodological strictness, arguing against artistic research ever really being research. If art involves artistic research, what about the underlying logic of the resulting knowledge, accountability, general validity and reliability of the approach adopted would be so different from what artists do automatically already? In such debates about artistic research, in short, a fundamental and simple difference is implicit between research that is associated with scientific strictness, trustworthiness and objectivity versus research that is associated with subjectivity, artistic autonomy and creativity. It is the reoccurrence, the dominance and the ideal-typical character of the implicit dichotomous yardstick that concerns me here. This reoccurrence is important,because it tells us something about the interests at stake. And to be sure, something is at stake when the autonomy of art, academic standards, or the relationships between academia and the arts are being put forward repeatedly.[3] However important these attempts to define and defend artistic research may be, they do not give us tools for getting a handle on the practice of artistic research, on what artistic researchers actually *do*, let alone help us to develop artistic research practices.

The too crude and ideal-typical character of the terms posited to define artistic research are, on the other hand, sometimes absent in actual *cases* of artistic research, in which the opposite problem is apparent. Here, instead, the focus is often largely on a factual recounting of the research process itself. See for example this opening paragraph of an article reporting on research related to scenography:

> This article is a first attempt to grasp our way of working and the intentions and assumptions that support it. I will pay attention to the strategies we use, the kind of effects these evoke, and start to find the right words and concepts to critically describe and reflect on a method that is neither fixed nor finished, and that has been developed and still is developing in and through our activities. (Merx, 2018, para. 2)

The author then explains why the article stays so close to the matter at hand and why this description of method is not theoretically or historically positioned:

> It is not the aim of this article to contextualize our strategies historically or theoretically. Although such framing could definitely be interesting and valuable – for example, how does our way of working relate to the artistic interventions in public space of the Situationists or to Henri Lefebvre's theory of the production of space – it would entail a framing after the fact. These theoretical or historical references did not play a role in the development of our project at the time. And our interest now is not so much to establish a context around our work, but rather to unfold our work from within, and to describe and analyse as precisely and accurately as possible our methodology, that up till this point has remained rather implicit and intuitive and is, as we believe, worth unpacking. (Merx, 2018, para. 3)

The importance of such detailed and specific work to stay close to the matter at hand and to be faithful to the procedure of unfolding research itself is not to be underestimated, particularly for an unfolding area of research such as artistic research. However, it also evokes the idiosyncrasy of individual projects and begs the question how (critically) to relate cases to other, possibly relevant, projects. (See other examples of specific artistic research publications in, for instance, the journals http://jar-online.net, http://forum-online.be, or *Art/Research International*, to name but a few.)

In this chapter, I want to try and fill the gap between the ideal-typical character and dichotomous terms of the demarcation debates about artistic research, on the one hand, and the scarcity of appropriate language to discuss and position artistic research in specific instances, on the other. In the following paragraphs, I will therefore look for a language that helps us to deal with questions like: What is it that artistic researchers do? What actually are the outcomes of artistic research? And what are good words to speak about what artistic researchers do and make? To do so, I will turn away from (specific examples of and debates about) artistic research and conduct a thought experiment. I will turn to two studies that together provide me with a model for the practice of artistic research. First, I will look at work by Stefan Hirschauer, who analyses fundamentally what it means to do ethnographic work and helps to find a productive middle ground between subjectivity and objectivity, between unruliness and methodological order. Secondly, I will focus on a piece by artist Pilvi Takala, whose work also can be seen as close to anthropological practice but who adds a risky, interventionist, active maker's perspective.

It is this area in which art and anthropology as well as description and intervention meet that I believe to be productive for developing a language to speak about artistic research. As part of my thought experiment on artistic research, then, I look for inspiration in ethnography/anthropology. This is hardly an uncommon thing to do.

For years, the affinity between artists and anthropologists or ethnographers has been noticed and analyzed, and sometimes criticized (see, for instance, Foster, 1995; Rutten, Van Dienderen & Soetaert, 2013; Schneider & Wright, 2006; Wright & Schneider, 2010; imaginativeethnography.org.). Likewise, the relation between performance studies and anthropology is close (see, for instance, for classic studies, Conquergood, 1992; Turner, 1986). This thought experiment draws on this work, with a keen interest in the actual, practical work that goes on at the borders of these disciplines.

In this chapter, I will look very closely at the two cases "as if" they were artistic research and see what can be gained from such an exercise. I do so by deliberately misreading or misplacing both examples. Hirschauer will be read more like an artist, while Takala will be understood more as an anthropologist or sociologist. Together, both examples suggest non-reductive ways in which we can grasp both the strictness of emerging methods as well as the space for that which escapes such methods, in academic as well as in artistic work. In conclusion, I will come back to the question of how the debate about artistic research has benefited from this STS move.

Writing

For the beginning of an answer to the questions above, we will listen to Stefan Hirschauer whose work I will discuss extensively here. He is interested in the research practice of ethnographers (being one himself). Ethnographers study the everydayness of (strange) cultures, often by means of participant observation and fieldwork, immersing themselves in and trying to become part of a culture. Hirschauer focuses on ethnographic researchers who do not simply fall back on tested and standardized methods, but who do not simply fall back on themselves as they are either.[4] He writes about writing. His analysis of the role of writing in ethnography will serve as a first model of what we might understand the practice of artistic research to be as well as providing words to speak productively about this practice.

In the article "Putting Things into Words: Ethnographic Description and the Silence of the Social" (2006), Hirschauer begins with the straightforward observation that ethnographers write. They not only write books and articles in which they report their research, their research itself involves lots of writing – the archetypical image of the ethnographer shows her bent over her notebook during fieldwork. There are two standard ways to account for this kind of writing, Hirschauer says. First, writing is meant to counter the researcher's forgetfulness. Fieldwork notes make sure that events and things that inevitably have an ephemeral character are preserved. Second, writing provides authority to the ethnographer. By writing, the ethnographer can justifiably claim: "*I was there* and what I have to say isn't nonsense."[5]

However, the rise of recording technology – from the voice recorder via camcorders to the iPhone – has had great influence, according to Hirschauer, on the thinking about the role of writing in ethnography and in the social sciences more generally.[6] Recording technologies seem to cut unmediated bits from reality in order subsequently to save them unaltered. Also, more nuanced thoughts on the value and impact of recordings by comparison turn fieldwork notes into merely diluted extracts. The researcher with her pen and notebook cannot write fast enough to keep up with the world around her, and her gaze is inevitably colored. As an ideal, in short, recording technology helps us to see what ethnographic description *cannot* do. But what, Hirschauer wonders, can it do?

To explain this, Hirschauer introduces interviews. The interview is a favorite tool within the social sciences for gathering knowledge. Unlike the interviewer, he says, the ethnographer by definition finds herself in the position of the observer. For the interviewer, language is the main source of information. But there is a problem if the interviewee falls silent. This problem serves as starting-point for the observing ethnographer, who is precisely interested in the things often left undiscussed, "material settings, wordless everyday practices, silent work practices, visual performativity etc." (Hirschauer, 2006, p. 423). It is her task to try to speak about "anything that happens mutely, as a wordless, inarticulate, 'illiterate' process" (ibid., p. 415). The ethnographer observes silence.

Hirschauer talks about "the silence of the social" (2006, p. 423) and distinguishes different kinds of silence. Silence can be about taciturnity and muteness, about those who have no voice, about speechlessness, the indescribable, the pre-linguistic, implicit knowledge and the self-evident. Sometimes it is about taboos that make it impossible to speak but also about verbal hesitations. He talks about that which for members of a community is so natural that it does not even need to be put into words, as well as about the powerless who have no opportunity to raise their voice. It is about the quality of the material world, ephemeral experiences or feelings that are hard to articulate. All these forms of silence are beyond description – because of their complexity, polyvalence, fleeting nature and so on. The metaphor of silence points to "a mute challenge for description to 'make something speak' that resists verbalization" (ibid., p. 423 n. 13). To that which in research limited to language escapes the researcher's attention, or which "stirs", as it were, as soon as the researcher tries to put it into words – as some sort of defiance of the "void".

How does the ethnographer deal with that void? To speak of that which does not show up automatically in language, Hirschauer says that the ethnographer takes herself seriously as observing research instrument. The association of observation with seeing from a distance is misplaced here.[7] The ethnographer in fact capitalizes on her connectedness with the world. The non-verbal aspects of reality are preeminently the domain of subjectivity, of the body and the senses. On account of her particular and precise preparation, the body of the ethnographer is receptive to the world of which she is part. For this reason, observation is always self-observation as well: the ethnographer can detect – smell, taste, feel, perceive – something in the world she studies by paying attention to what she herself experiences (Hirschauer, 2006, p. 426). But how does such attention emerge? What does the ethnographer *do* to detect things in an informed and skillful way as research instrument?[8] Hirschauer's reply is: writing. Writing is the researcher's specific cultivation technique or "discipline of subjectivity."[9]

What Hirschauer mainly addresses here is the writing performed by the ethnographer during fieldwork. Writing is a technique that both enables and sustains the distancing an ethnographer needs. If the alternation of strange and familiar is typical of anthropological fieldwork, it is also necessary to gain access at all to that which does not easily reveal itself. "The most powerful recourse for dealing with things taken for granted ... is the unfamiliarity of the observer" (Hirschauer, 2006, p. 433). First impressions are crucial. By making notes the ethnographer may constantly recharge that position. She can thus vary the distance and actively produce her own estrangement (ibid., p. 433). How does this work? Why is writing so fitting for creating distance? Writing allows the ethnographer momentarily to withdraw from the field: "Just arrived, one already retreats again into a highly self-absorbed activity" (ibid., p. 427). Retreated with her notebook, she produces a specific kind of *attention* by recording notes: "the compulsion to write facilitates a mnemonic state of awareness" (ibid., p. 416).

This opens up space and the possibility of an analytic perspective on the field. Reading and rereading notes, in the field but also later, strengthens this perspective. It is reinforced by the fact that writing and rereading require time and concentration. "Memorizing by writing is a kind of 'rumination' " (ibid., p. 428). Writing causes our memory to be active and our memories to last longer. Writing, in other words, solidifies time by disciplining our memory (ibid., p. 429). But prior to that, writing already disciplines the ethnographer's gaze. Just as the awareness that later on you will have to start writing again generates attention now for specific details.

Although notes *describe* things as well, Hirschauer stresses the constructive character of descriptions (2006, p. 437). And although anthropological handbooks often emphasize the need to separate observation from interpretation and analysis in fieldwork reports, Hirschauer argues that recording notes enables the continuity and connection between observation, description and analysis (ibid., p. 430).[10] Moreover, describing also has an explicitly creative element here. At moments at which the researcher cannot yet find any words herself, she may experiment: she can try out different descriptions, for instance by using language from elsewhere. Or she can deliberately use language that does not apply or consciously take too literally language she has taken from the field (ibid., pp. 433–434). The ongoing challenge is to search for the right words – to grope for them, taste them.[11]

> Describers, struggling with their own speechlessness are less in the situation of a narrator, retrospectively reconstructing a subject matter, and more like a taster, who has to use categories situatively and deliver them to others. The words they are looking for are on the tip of their tongue – just like a wine one is trying to describe. (Hirschauer, 2006, p. 427)

A description can function as a constructed lens "to 'look *through*' in order to gain new ways of seeing" (ibid., p. 433).

So, what does Hirschauer's understanding of the role of writing in ethnography change how we might speak about doing artistic research? We no longer understand observations of artistic researchers to automatically produce standardized, reliable descriptions, nor do we understand them only as subjective or as appearing as by magic from within the observer. Rather, observing involves a process of intense self-observation by the embodied researcher, that can result in a greater sensitivity for the object of study. This object can only make an impression on the researcher-as-instrument through her attitude of active and creative construction. Such an attitude emerges in the sensitive subjectivity pursued in the preparatory stage and the sustained discipline of writing, with which the ethnographer is groping for words to mitigate the silence.

Even without defining the ethnographer's work from the ideal of recording technology, the ethnographer with her notebook who is left to her own devices appears rather helpless. It is embarrassing, Hirschauer says: "As field researchers they stand alone with their experiences, at a loss for words .... Even worse, they find themselves compelled to be their own source .... What a frightening amount of freedom!" (2006, p. 423). That which the ethnographer writes is necessarily "homemade" (ibid., p. 437), but this is not to suggest that anything goes for ethnographers. On the contrary, if writing is constructive and creative, Hirschauer is clearly concerned with *de*scribing. In all her embarrassing underdetermined self-cultivation, the ethnographer is fully geared to creating sensitivity for the world she studies. She neither just simply *does* something nor engages in empty navel-staring. For Hirschauer, it is constantly about that which is enabled by self-observation and describing: "Giving the field ... a maximum of opportunities to 'inscribe' itself into researchers and author" (ibid., p. 437).[12]

Hirschauer shows the need for and the opportunity of this kind of descriptions, which he calls "high-risk" (1994, p. 339). By doing so, he opposes an understanding of ethnographic description as being about faithful representation. Risky interpretations have value because they were made by a sensitive research instrument that combines intimate and intensive contact from within with distance and analysis from the angle of another world. As a result, descriptions emerge that necessarily do *not* fully coincide with the field. The

> verbalization of the silent does not come without a basic transformation of the object. The social does not exist in language on its own. *This* (and not the adding of an observer's interpretation) is the decisive factor why descriptions cannot be doubles. (Hirschauer, 2006, p. 436)

And:

> Opening up forms of silence always also implies a break with the membership in the field: betraying loyalties to the unmentionable, overcoming intense experiences that render us speechless, breaking with participation competence, which keeps knowledge "*tacit*", and formulating what is taken for granted and thus in need of no words. (Hirschauer, 2006, p. 436)

Too many descriptions, Hirschauer argues, aim for faithful representation of the world. But today we seem to have too many versions of reality already that merely reproduce that reality. As a result, a much more important problem of contemporary anthropological research is pushed to the background.[13] Such descriptions "suffer from a *lack of distortion*" (Hirschauer, 2006, p. 438). What we need are "those de-contextualization processes in which a perspective is created, a vision of the field that differs starkly from the way members see it" (ibid., pp. 438–439). To make descriptions that *deviate* or *distort*, researchers have to make use of the embarrassing freedom they have. And this freedom they need to discipline again with pen and paper. "Descriptions have to *make use* of their liberties. But this also means that they will be judged according to their *analytical* achievements" (ibid., p. 439).

The self-observing, tasting, probing, writing and rereading ethnographer will take on this task. The alienating possibility of the researcher's saying different things about the world will precisely emerge in the space between what Hirschauer calls the "silence" of the world and the verbal means of the ethnographer. Hirschauer's analysis thus provides an elegant detour that helps to escape the clichéd dichotomy between artistic freedom and academic method. But it also provides a vocabulary to discuss and position specific exemplars of artistic research. The artistic researcher is not confined to a set methodology nor at liberty to do whatever she likes. She is a crafty, self-cultivated writer of silence.

But is that really what I want to say? Is *writing* what artistic researchers do? Don't artists usually do rather different things than writing? Things like rehearsing, designing, making, performing et cetera? And aren't such activities a much better source for the kind of risky descriptions of silence Hirschauer is after, since they often express exactly that which cannot readily be put into words? Hirschauer also hints at the issue of whether other means than language are more suitable for studying silence. Language has its limitations of course – limits he takes for granted in his article (Hirschauer, 2006, p. 436), but:

> Outside the social sciences, language is of course transgressed in almost all the arts, especially in music; literature too has a "zone of the ineffable" .... Within sociology, the possibilities are more limited, but nevertheless, interesting: in "ethnodramatic" performances of research results ... or in attempts to use visual media not only for the recording of "data," but also as a semiotic extension of sociological communication, and as an analytically articulate form of showing. (Hirschauer, 2006, p. 437 n. 28)

Following Hirschauer here, art is perhaps the domain par excellence where we expect to see ways of working that transcend the verbal. Artists do all sorts of things, but in most cases writing and reading do not top their list of activities.[14] Let us take Hirschauer's ideas about the writing ethnographer along with us as we turn to the second example in my thought experiment. Let us step with Finnish artist Pilvi Takala into an elevator. In her "subtly transgressive video performances" Takala infiltrates "ambiguous and exclusive communities" like the marketing department of Deloitte company (Lange, 2012, para. 1).

Spending an Entire Day in an Elevator

Pilvi Takala's website (pilvitakala.com) contains a list of works. One of them is called *The Trainee* (2008). For this project the artist worked for a month as the intern Johanna Takala in the marketing department of Deloitte company.[15] The work consists of an installation that includes a Powerpoint presentation, a letter and several videos. A few fragments are on the website, including "February 28, a Day in the Elevator" (Takala, 2008, 06:36–13:00 mins). In this fragment, the intern is standing in an elevator within the company. The camera, held quite low, is probably hidden, perhaps in Takala's bag. We do not see her, but look from her position into the elevator. The footage is shaky and out of focus, suggesting that this is a recording of the real world. What we seem to be seeing is real life, not so far away from our own lives. We see people entering and leaving the elevator, and we overhear (as I don't understand Finnish, by reading the English subtitles) the emerging conversations: Which floor are you going? Are you still here? What are you doing here? The series of brief fragments end with an e-mail and with the following message recorded on an answering machine:

> I just wanted to say that your trainee has found a new place. So now she is lodging in the elevator ... and she's standing in the elevator and ... rides back and forth with people. I mean she doesn't press the button herself, but just stands there in the corner. There's such problems involved in here that I don't want to go into the same elevator with her. So, she's riding with the elevator back and forth. So try to get her out. Ok, bye! (Takala, 2008, 13:02–13:32 mins)

In other short films we see the trainee at an empty desk in an office or just staring intently at nothing while sitting in the company library, not just briefly but for a long, long time. In these corporate places, it is highly uncommon and easily noticed when someone is doing nothing. As viewers we also *see* this behavior being at odds with the setting.

Colleagues sitting nearby work in a concentrated fashion behind their PCs, are on the phone loudly and walk away confidently and self-assured, until their gaze catches Takala. We see them respond to the passive trainee in increasingly concerned or agitated ways. Her seemingly unconcerned and sustained inaction magnifies how *work* is normally performed there. Moreover, the tone on the answering machine, the exclamation marks in the e-mails, the increasingly awkward questions and the hesitating steps in the office underscore the importance attached to the breached norm.

Watching this short film made me curious to know what took place right before and after it. How did Takala arrive at doing nothing in the elevator? Where did she get the notion but also, and in particular, what did she do, see, think, feel or contemplate to get there? And what in fact happens in that elevator? What kind of situation is it that emerges there? What are we as viewers, who watch this video on the internet, to do with it? And what happened later on in the company in response to her uncommon presence? How, in other words, did Takala become a sensitive ethnographer à la Hirschauer? And what does her work add to his? How can we understand Takala's work in such a way as to provide a model of what artistic research might be in practice?

To understand Takala's video projects "as if" it were artistic research, ethnomethodologist Harold Garfinkel's notion of a "breaching experiment" is helpful.[16] In the 1960s Garfinkel developed his famous breaching experiments. A breaching experiment (literally: an experiment in violation) is a way of uncovering implicit norms that have general validity in a community. In the original version of the experiment, Garfinkel gave his students different assignments. For instance, they had to imagine that they doubted everything someone said. In some cases, they also had to act on their imagined role. The best-known assignment was that students had to behave at home as someone renting a room there (Garfinkel, 1963–1964). Garfinkel did not find these experiments to be real ones. "They are demonstrations, designed, in Herbert Spiegelberg's phrase, as 'aids to a sluggish imagination' " (ibid., p. 227). It is a tricky thing, one might say, to capture the ordinary course of affairs – life as it happens every day, the things we do not discuss or mention or not even really perceive. Even if you would want to, how do you perceive it? By definition, after all, it involves that which you do not know, which is taken for granted, which you assume. This is, in other words, about Hirschauer's silence. Garfinkel's smart innovation – and this may sound banal – is to get going and see what happens. The everyday reality will resist, he seems to suggest. "Procedurally", Garfinkel says, "it is my preference to start with familiar scenes and ask what can be done to make trouble" (ibid., p. 227).

Garfinkel has a clear idea about the kind of *trouble* he is after. He tries to evoke feelings of confusion, stress, indignation, fear and embarrassment (Garfinkel, 1963–1964, p. 227). He has good reason for doing so. The confusion and the emotions that emerge in the experiments "produce reflections through which the strangeness of an obstinately familiar world can be detected" (ibid.). Strangers will immediately see what we experience as familiar, and, if not, it actively needs to be made strange to us (ibid., p. 226).

Takala's project at Deloitte may be viewed as a breaching experiment. If her interventions, as we saw, caused us to become estranged from the familiar world of working hard and being active, even more seemed to be going on in the elevator.[17] There is hardly any room to let your gaze roam freely or do anything really. Moreover, the elevator fixes you mercilessly in that situation for the duration of the ride. All you can do in a conventional elevator is look, without having any options to assure your fellow elevator users that you do not want anything from them. This sounds perhaps like an unusual situation, but, as Hirschauer argues:

> Unknown others become strange because of a radical devaluation of physical co-presence as a chance for establishing contact. Urbanisation and geographical mobility have inflated this chance by producing masses of "*insignificant others*," clashing in encounters without past and future. Personal relations only gain their meaning against this background of a massive unrelatedness. (Hirschauer, 2005, p. 59)

The strangeness we produce in an elevator resonates with the modern urban experience, in which we constantly work to maintain a kind of nothingness and non-relation. Estrangement, conceived as the production and sustenance of absence, is in fact normal in the elevator.

A breaching experiment has the same goal as the ethnographer's writing: to make the familiar strange. But alienation does not emerge because the researcher *withdraws* from the world of which she became part in order to capture it in other words. Alienation emerges *in* and *with the help of* reality. This alienation has a disclosing effect. By empathizing with the people in that elevator, the trainee and the employees, we as viewers see their implied normal world with fresh eyes but also that of ourselves – the world perhaps in which we also say nothing in the elevator, the world in which we work hard at our job, the world in which we would look upon someone who has the luxury to do nothing at her job with envy.

What does Takala's work become when seen through the lens of Garfinkel's breaching experiment, as well as following on the ethnographer's work as described by Hirschauer? As Hirschauer describes, Takala needs to coincide with the world sensitively. This is an important precondition in order subsequently to be able to cause problems. But in this case her work does not involve writing but poking and teasing in the world as it is. Instead of looking for distorting words, Takala distorts the world and our experience of it. Either way, and regardless of language or world, the effect emerges once we turn back our gaze upon normalness. *It* takes on a new, fresh quality.

Takala's elevator is actually a kind of pressure cooker that raises the temperature of the normal alienation and absence that we encounter on a daily basis. Her passivity in the elevator is a breaching experiment that breaks the implicit norm of the activity we associate with labor. Her nothingness is not masked or veiled, and in her films we see and hear the uneasiness and stammering it causes: contact, conversations, e-mails, voicemails. This uneasiness, however, is an effect of breaking not only the norms commonly tied to *work* but also the norms commonly tied to elevator behavior. Although in the elevator Takala is passive, an uneasy sort of contact does emerge. With her somewhat vague, open reactions and her mild questioning tone, she seems to actually pursue having such conversations and having them continue and develop. Accordingly, if we apply Hirschauer's analysis to the elevator experiments, Takala's work also points to another fundamental quality of our everyday, normal life: one in which the absence of contact has to be sustained in discreet ways.

Instead of evident uneasiness and thus the exposure of an implicit normality, Takala produces something richer, something which is more ambiguous and lasts longer.

Different norms converge here and diametrically highlight each other. What from one perspective (work) is odd (passivity) is familiar from the other perspective (elevator), and vice versa. This opens up the possibility of something beyond unease. Takala's project suggests a way to bypass the critical and exposing yet also rather blunt and one-dimensional effect of breaching experiments.[18] Moreover, Hirschauer's silence had changed character through this maneuver. In the case of Hirschauer, the silence was hidden in the world in which the ethnographer was present through observation and writing. Because this silence was not cut and dried, the ethnographer had to deploy her writing discipline in order to find other words for it. Likewise, in Garfinkel's breaching experiments, the silence is the object of study that escapes immediate attention. And here, too, the researcher, through participation, needs to become sensitive for the world in which this silence is hidden. But what about Takala? Instead of looking for words through writing, the researcher puts her finger on a spot that hurts and waits for what happens. She breaks the silence and listens attentively to resonance.

In Conclusion

Hirschauer described the ethnographer as someone in the position of a taster who, in her quest for an appropriate language to grasp that which perpetually escapes her, has to experiment with language. She may even need to turn elsewhere to transport and misuse language developed in other realms. In a way, my thought experiment in this chapter itself is an example of this. To try to find a language that helps us to speak about the doing of artistic research without reducing it either to specific cases or to clichéd imagery of what art or science might be, I turned to two works. In my thought experiment I deliberately pretended that the two works I studied were in fact instances of the practice of artistic research. I did this based on what I have learned from STS research: by staying very close to the material that I studied and to allow that material to infuse bottom-up my understanding of it. Moreover, I read Hirschauer a little like an artist and positioned Takala alongside academic researchers such as Garfinkel, even though they might never (and have not) done so themselves. Hirschauer provided words to understand the sensitive body of the researcher, that by skillfully sensing becomes part of the world. This body was then handed more interventionist, more exploratory and distorting possibilities by Takala and Garfinkel. Here the researcher does not need to detach herself or develop a distorting language or perspective. Rather, she first needs to *make* trouble, for real. Alienation will follow in its wake.

What does this STS inspired thought experiment yield? We have encountered a process in which the artistic researcher turns herself into an instrument. There is no set method nor is there artistic freedom as such. Rather, the instrument attunes itself and develops standards along the way. It mixes immersion with estrangement and observation with intervention. It produces risky descriptions of silence, as well as riskily staging silence. It works in a place somewhere between lab and reality, stage and fieldwork, knowing and undergoing. These experimental situations do not produce academic knowledge nor can they unequivocally be conceived as artists' work. They should fundamentally be understood, I would like to argue, to be perched on the brink of embodied knowledge production, on the one hand, and a kind of practical ethics, on the other: as disturbing and imaginative practical exercises, proposals, attempts.[19] Exercises, proposals and attempts that are difficult to place, and that fundamentally challenge us to determine what kind of registers – academic, artistic, ethical, aesthetic – should be applied to them.

This chapter has the aim of getting away from dichotomous and clichéd definitions of artistic research. Such definitions may be good for defense or attack but are unfit for the more mundane and difficult task of practicing the craft of artistic research. Based on my "as if" reading of Hirschauer and Takala, I would argue that artistic research has silence as its core object of study and that only via the fierce and meticulous study of silence does it come to be reflexive about what art is but also about what counts as knowledge and ways of creating it and making it count. This also means that rather than using artistic research in a renewed defense of art as an autonomous field – often an implicit thread in debates about artistic research – I see artistic research as risky interference with silence, which is by nature contextual and thus potentially critical. I hope that my concentration here on this space between ethnographic research and artistic practice will encourage practical exercises in artistic research and thus allow us to ask different questions about it. Questions like: What kinds of self-cultivation do artistic researchers use? How do they sensitize and calibrate themselves? What kinds of disturbing silence do they evoke? How do their risky interferences with silence relate to others? And how and where do they matter?

Notes

1 This chapter is a condensed and partial version of my inaugural lecture as head of a the Research Centre for Arts, Autonomy and the Public Sphere, at the Faculty of the Arts at Zuyd University of Applied Sciences in Maastricht, The Netherlands, see Benschop, 2015. Thanks for their support and useful comments: Jasper Coppes, Douwe Draaisma, Ruud Hendriks, Maaike Lauwaert and Peter Peters. The English was much improved by Ton Brouwers who translated the inaugural lecture.

2 For an overview of STS, see, for instance, Hackett, Amsterdamska, Lynch & Wajcman, 2008. On Actor–Network Theory specifically, see, for instance, Law & Hassard, 1999. Classic STS studies include Callon, 1986; Latour, 1987.

3 From the huge literature discussing the relationship between art and science, see, for instance, Boekmanstichting, 2004; Born & Barry 2010. Pascal Gielen has written a lot on the relations between cultural politics, repressive liberalism and the professional audit culture. A compact text is Gielen, 2014. For critical work on developments in today's university, see Halffman & Radder, 2013; the special issue of *Krisis* (2015). On the increased academic nature of art and artistic research, see, for example, Bears, 2011. Pels (2001) writes on the value of academic slowness.

4 I largely ignore here the debate on the differences or similarities between anthropology and ethnography. Ingold (2013) makes a distinction between ethnography as positivist practice in which the faithfulness of representation is center-stage, while anthropology is essentially aimed at developing attention as well as at transformation. If Hirschauer's view of ethnography, to be discussed in the next section, is empirical, it is transformative as well. For a discussion, see Coumans, Sonderen, Benschop & Strandvad, 2018.

5 Anthropologists and ethnographers do not just write during their fieldwork, they also publish their results. To write and speak about and in the name of others comes with other, political problems, which postcolonial anthropology struggles with and reflects upon. In postcolonialism the political and epistemological questions that can be asked about assumptions and effects of knowing the Other have come to be center-stage. For an introduction, see Young, 2003. The democratization of voices and practices and the questions it triggers about expertise, power and empowerment do not just feature in anthropology or in studies of culture but, much more broadly, also in cultural and artistic practices.

6 Hirschauer does two things in this article. He establishes recording technology as a standard, but he also looks at what the practice of recording itself produces. This also comprises

criticism of the ideal of recording. He concludes that both descriptions and recordings are sociological artifacts that shape data registration. The norm that emerged through the rise of technologies of registration also helps to better understand the nature of the specific quality of writing. Hirschauer also discusses the quality of recordings, which is another one than the ideal suggests: "Thus, the special quality of technical preservation does not reside so much in its capacity to make a literal copy, or in the neutral manufacture of a textual double, but in creating something entirely new: the singular, self-identical conversation" (Hirschauer, 2006, p. 420). His analysis reminds one of that of Auslander (1999), who argues that media such as television are a precondition for the idea of live, unmediated performances as such.

7 The literature on seeing and distance and proximity is enormous. I would like to refer here to the work of Ike Kamphof (2013).

8 On the researcher who calibrates herself in a quite different discipline to function as an effective instrument, see Benschop & Draaisma, 2000. For recent applications of this idea to artistic research, see Kroese, Spronck & Vermeulen, 2017; De Rooij, Benschop, Nahar & Yaron, 2018.

9 Hirschauer (2006, p. 426) borrows the concept of "disciplined subjectivity" from Wolff, 1999.

10 The idea that description on the one hand and analysis or interpretation on the other should be separated from each other is also backed up by the ideal of recording technology, in which storing empirical material and the interpretation of it do not take place at one moment or in one location (in the notebook) but in different media and at different moments. For sound reflection on the writing of fieldwork notes, see Emerson, Fretz & Shaw, 2011.

11 For great writing exercises, see work by Peter Elbow (1981), brought to my attention by Ruth Sonderegger. For reflections of authors on their writing practice, see the interviews in the *Paris Review* series "The Art of Fiction" (www.theparisreview.org) (I am grateful to Anna Harris for this reference).

12 The focus is here on what Ingold means (e.g. Ingold, 2013) when he describes anthropology as an "education of attention" or, rather, learning with and about the world.

13 The problem addressed here by Hirschauer is the superfluity of ethnographic description (2006, p. 438), given the proliferation of self-descriptions in the world. Traditionally, anthropologists viewed their role as preserving that which threatens to become extinct, a goal that grew problematic with the emergence of postcolonial thought and the democratization of knowledge. Rather than viewing this democratization as a problem, Hirschauer considers self-representations from the field that fully coincide with that field as such.

14 Unless they are writers themselves of course. Moreover, in particular, painters have a long tradition of writing about their work: just think of the letters by Van Gogh or Sol LeWitt. Also in other ways, artists, directors, musicians, architects, performers of course often work with and through writing. Here too, it is often the non-representational character of language that is important. However, I by no means want to suggest a fundamental divide between academia and the arts along these lines. I am more interested in the various ways that writing is used as a tool for sensitization – both in academia and in the arts. I am grateful to Peter Peters for bringing up this issue.

15 "Deloitte is one of the world's leading professional companies. We operate in over 150 countries providing audit, consulting, financial advisory, risk management, tax and related services to select clients. With the help of Deloitte's international network and strong local know-how we are able to serve our clients comprehensive, innovative solutions and help them succeed in their business. In Finland there are 550 of us and globally over 286 000" (Deloitte, 2019).

16 My thanks go out to Vivian van Saaze, with whom I intensively reflected on breaching experiments, and Tino Seghal. I also think back to a workshop during Artful Encounters (http://virtualknowledgestudio.nl) in which Ike Kamphof and Anne Beaulieu posed critical questions about this kind of intervention. For a critical analysis of breaching experiments, see Pleasants, 2002, ch. 7.

17 I base the analysis of the elevator as laboratory for research of the production of non-relations on Hirschauer, 2005.

18 A great example of interventions that tamper with implicit expectations while at the same time, as in Takala's work, design a situation in a way that it shifts and lasts, are those by

Antanas Mockus. As mayor in Bogota, where traffic violations were a big problem, he had traffic regulators be trained as mime artists. See, for example, VPRO, 2010.

19 For the shift of thinking about artistic research in terms of practical ethics rather than in terms of knowledge practices, I am indebted to Ties van de Werff.

References

Auslander, P. (1999). *Liveness: Performance in a Mediatized Culture.* London and New York: Routledge.

Bears, M. (2011). Inside the Box: Notes From Within the European Artistic Research Debate. *e-flux journal.* Retrieved from https://www.e-flux.com.

Benschop, R. (2014). Practicing the Artistic Research Catalogue. In M. Schwab & H. Borgdorff (Eds.), *The Exposition of Artistic Research. Publishing Art in Academia* (pp. 105–117). Leiden: Leiden University Press.

Benschop, R. (2015). *De eland is eigenwijs dier: Een gedachtenexperiment over praktijk en relevantie van artistiek onderzoek* [inaugural lecture]; Lectoraat Autonomie en Openbaarheid in de Kunsten: Zuyd. Retrieved from http://lectoraataok.nl/wp-content/uploads/2016/06/Inaugurele-rede-Ruth-Benschop.pdf.

Benschop, R. & Draaisma, D. (2000). In Pursuit of Precision: The Calibration of Minds and Machines in Late Nineteenth-century Psychology, *Annals of Science, 57*(1), 1–25

Biggs, M., & Karlsson, H. (Eds.) (2010). *The Routledge Companion to Research in the Arts.* New York: Routledge.

Boekmanstichting (2004). *Kunst en Wetenschap.* Special issue of *Boekman: Tijdschrift voor Kunst, Cultuur en Beleid, 58/59.*

Borgdorff, H. (2012). *The Conflict of the Faculties: Perspectives on Artistic Research and Academia.* Leiden: Leiden University Press.

Born, G., & Barry, A. (2010). Art-Science. *Journal of Cultural Economy, 3(1)*, 103–119.

Callon, M. (1986). Some Elements of a Sociology of Translation: Domestication of the Scallops and the Fishermen of St. Brieuc Bay. In J. Law (Ed.), *Power, Action, and Belief: A New Sociology of Knowledge* (pp. 196–223). London: Routledge & Kegan Paul.

Conquergood, D. (1992). Ethnography, Rhetoric, and Performance. *Quarterly Journal of Speech, 78*, 80–123.

Coumans, A., Sonderen, P., Benschop, R., & Strandvad, S. M. (2018). Dwell / Act / Transform: Three Views on an Exhibition of Artistic Research. *FORUM+*. Retrieved from http://www.forum-online.be/nummers/zomer-2018.

De Rooij, M., Benschop, R., Nahar, C., & Yaron, G. (2018). *Inter/Face: Reflecties op een onderzoeksatelier.* Maastricht: Lectoraat Autonomie en Openbaarheid in de Kunsten, Maastricht Academy of Fine Arts and Design, Zuyd Hogeschool.

Deloitte (2019). Who We Are, and What We Do. Retrieved from https://www2.deloitte.com/fi/fi/pages/about-deloitte/articles/who-we-are-deloitte-finland-about-deloitte.html.

Elbow, P. (1981). *Writing with Power.* New York: Oxford University Press.

Emerson, R. M., Fretz, R. I., & Shaw, L. L. (2011). *Writing Ethnographic Fieldnotes*, 2nd ed. Chicago and London: University of Chicago Press.

Foster, H. (1995). The Artist as Ethnographer? In G. Marcus & F. Myers (Eds.), *The Traffic in Culture: Refiguring Art and Anthropology* (pp. 302–309). Berkeley: University of California Press.

Garfinkel, H. (1963–1964). Studies of the Routine Grounds of Everyday Activities. *Social Problems, 11*(3), 225–250.

Gielen, P. (2014). Repressief Liberalisme: Over kunst, markt en cultuurbeleid in Nederland. In Lanoo Campus, *Starten in een creatieve loopbaan* (pp. 172–193). Leuven: Lanoo.

Hackett, E. J., Amsterdamska, O., Lynch, M., & Wajcman, J. (2008). *The Handbook of Science and Technology Studies*, 3rd ed. Cambridge, MA: MIT Press.

Halffman, W., & Radder, H. (2013). Het academisch manifest: Van een bezette naar een publieke universiteit. *Krisis: Tijdschrift voor actuele filosofie, 3*, 2–18.
Hannula, M., Suoranta, J., & Vadén, T. (2005). *Artistic Research: Theories, Methods and Practices*. Helsinki: Academy of Fine Arts Helsinki.
Hirschauer, S. (1994). Towards a Methodology of Investigations into the Strangeness of One's Own Culture. *Social Studies of Science 24*, 335–346
Hirschauer, S. (2005). On Doing Being a Stranger: The Practical Constitution of Civil Inattention. *Journal for the Theory of Social Behavior, 35*(1), 41–67.
Hirschauer, S. (2006). Putting Things into Words: Ethnographic Description and the Silence of the Social. *Human Studies, 29*(4), 413–441.
Ingold, T. (2013). *Making: Anthropology, Archaeology, Art and Architecture*. London: Routledge.
Kamphof, I. (2013). *Iedereen Voyeur: Kijken en bekeken worden in de 21e eeuw*. Zoetermeer: Klement.
Klein, J. (2012). What is Artistic Research? *Research Catalogue*. Retrieved from https://www.researchcatalogue.net/view/15292/15293/0/0.
Kroese, R., Spronck, V., & Vermeulen, M. (2017). *Drawing Instruments, Or: How to Calibrate an Artist?* Retrieved from http://lectoraataok.nl/nieuws/symposium-drawing-instruments-or-how-to-calibrate-an-artist.
Lange, C. (2012). In Focus: Pilvi Takala. *Frieze, 147*. Retrieved from https://frieze.com/article/focus-pilvi-takala.
Latour, B. (1987). *Science in Action*. Cambridge, MA: Harvard University Press.
Law, J., & Hassard, J. (1999). *Actor Network Theory and After*. Oxford: Blackwell.
Merx, S. (2018). Between Realities #Athens: Or How Scenography Can Facilitate the Re-imagination of Public Space. *FORUM+*. Retrieved from http://forum-online.be/nummers/zomer-2018/between-realities-athens-or-how-scenography-can-facilitate-the-re-imagination-of-public-space.
Nelson, R. (Ed.) (2013). *Practice as Research in the Arts: Principles, Protocols, Pedagogies, Resistances*. Basingstoke, Hants.: Palgrave Macmillan.
Pels, D. (2001). Wetenschap als onthaasting: En onthaasting van de wetenschap. *Krisis: tijdschrift voor empirische filosofie, 2*, 6–25.
Pleasants, N. (2002). *Wittgenstein and the Idea of a Critical Social Theory: A Critique of Giddens, Habermas and Bhaskar*. London: Routledge.
Rutten, K., Van Dienderen, A., & Soetaert, R. (Eds.) (2013). *Revisiting the Ethnographic Turn in Contemporary Art*. Special issue of *Critical Arts, 27*(5–6).
Schneider, A. & Wright, C. (Eds.) (2006). *Contemporary Art and Anthropology*. Oxford: Berg.
Slager, H. (2004). Kunst en methode. *Boekman tijdschrift voor kunst, cultuur en beleid, 58/59*, 197–202.
Spronck, V. (2016). "Between Art and Academia: A Study of the Practice of Third Cycle Artistic Research" (unpublished master's theses). Maastricht University.
Takala, P. (2008). *The Trainee*. Retrieved from http://pilvitakala.com/the-trainee.
Turner, V. (1986). *The Anthropology of Performance*. New York: PAJ Publications.
VPRO (2010). *Wintergasten: Raoul Heertje interviewt Antanas Mockus* [TV interview], 27 Dec.
Wilson, N., & van Ruiten, S. (Eds.) (2013). *SHARE Handbook for Artistic Research Education*. Amsterdam: Valand Academy.
Wolff, S. (1999). Subjektivität für alle praktischen Zwecke. *Österreichische Zeitschrift für Soziologie, 24*, 5–24.
Wright, C., & Schneider, A. (2010). *Between Art and Anthropology: Contemporary Ethnographic Practice*. New York: Berg.
Young, R. J. C. (2003). *Postcolonialism: A Very Short Introduction*. Oxford: Oxford University Press.

5 Wisdom in Artistic Research

An Alternative to the Discourse of Art as Knowledge Production

Nora S. Vaage

Introduction

In recent years, an increasing number of artists are engaging with the biotechnosciences, entering scientific laboratories to create art *in vivo*. The artists creating these living artworks are often concerned with the material and sensory nature of the bacteria, tissue, plant or animal with which they are working, as well as with the critical potential of their practice in relation to scientific institutions and methods. Their approach also entails a preoccupation with artistic research and the process of creation as endeavours of knowledge production. In this chapter I will take these bioart practices as a starting point for discussing the discourse of knowledge production accompanying artistic research, drawing on theories from science and technology studies (STS) and philosophy. In what sense of the word can we speak about artistic lab practices as producing knowledge? It has often been argued that few artworks can be seen to fulfil the criteria of expressing "justified true belief", which is still a common definition of knowledge within epistemology. According to this theory, we can be said to have knowledge when something is true, and we are justified in believing that it is true (Gettier, 1963; Klein, 1971). But this theory is seen, by several scholars, as being unsuitable for the ideas about knowledge entailed in artistic research. These scholars argue that attempts at fulfilling such criteria reduce art to science communication or exploit it in the interests of capitalist society (Busch, 2009; Steyerl, 2010). However, considering the contributions that artists make and building on critical ideas about knowledge and artistic research, I argue that a more suitable concept to apply to the meaning-making of art may be wisdom.

"Artistic research" has in recent years become a controversial, yet popular idea, with multiple interpretations. While we can argue that the process of creating art has always entailed some research, starting in the past decade the term is commonly used to describe artists' work within academic settings or in close collaboration with academics. The same tendency is also referred to as practice-led research, art-based research, artistic enquiry and arts-informed practices (Borgdorff, 2010; Mäkelä et al., 2011; McNiff, 2013; Smith, 2008). Programmes for practice-based PhDs and artist residencies have been established at universities and colleges all over the world and an increasing number of artists spend most of their time in academia, publishing papers and attending conferences as part of, as well as supplement to, their artistic practice. As suggested by Schaffer (2011), in addition to specific practices, this can also be seen as an institutional discourse, determined by recent funding schemes and incentives for artists to "do research". Efforts to develop artistic research programmes are often met with resistance by academics, for diverse reasons that range from concerns with the quality of the research outcomes to the argument that art cannot produce knowledge outcomes and therefore should not be defined as research (Garneau, 2008). In this latter view are inherent certain ideas about what research and knowledge is and should be that are quite different from the perspectives within artistic research. The unwrapping of the contested terrain occupied by these conflicting ideas of knowledge in artistic research will be attempted in the first part of this chapter. I argue that, despite alternative accounts of knowledge being widespread within STS and related fields, established ideas about academic knowledge shape the creation as well as reception of artistic research in ways that are not always fruitful.

As argued by McNiff, "there is a striking disconnect when people and professions understanding the unique resources of the arts in addressing problems and issues inaccessible to verbal analysis nevertheless persist in the use of the latter as the exclusive foundation of research" (2013, p. xiii). I will argue that a major reason for this persistence is the predominance of the propositional view of knowledge, in which knowledge consists of verbal, logical and truthful, propositions, in conceptions of (scientific) research. This predominance continues despite efforts by twentieth-century philosophers such as Heidegger, Gadamer and Feyerabend and, in the past few decades, STS researchers from Latour to Nowotny, to expand and complicate ideas about what (relevant) knowledge can be. While there is not space within these pages to draw up a comprehensive history of these arguments, I will focus on a few specific STS challenges to the propositional view of knowledge. From there, I will address ideas about knowledge within artistic research. Discussions about artistic research (with the notable exception of artists discussing their own research practice) have so far often remained at a rather theoretical and overarching level (see e.g. Busch, 2009; Mäkelä et al., 2011). This carries in it a danger of assuming that all art forms have similar processes and challenges in their research endeavours. To avoid this pitfall, and to showcase some of the empirical material on which these ruminations are based, I will consider some examples of the research processes and outcomes of artists working in bioscientific laboratories. Given its unique status as the only permanent artistic research centre situated within a biology department, I will focus my discussion on observations made during a three-month case study in 2013 at the SymbioticA Centre for Excellence in Biological Arts, in particular discussing two artistic projects by then-artists-in-residence Benjamin Forster and Nigel Helyer.

Conceptions of Knowledge

What is knowledge? This question has been posed again and again throughout the ages, and the answer is dependent on societal, historical and disciplinary context. We live in a knowledge society, famously, but what do we understand knowledge to be? What forms of understanding, data, and skillsets are included in our concept, and what others, consequently, are excluded?

In Chorus I of his poem "Choruses from The Rock", T. S. Eliot asks

> Where is the wisdom we have lost in knowledge?
> Where is the knowledge we have lost in information? (Eliot, 1963, p. 147)

A few lines earlier he claims: "All our knowledge brings us nearer to our ignorance, All our ignorance brings us nearer to death". In our contemporary knowledge society, where information is ubiquitous and the race for technological development gains speed, this poetic association of knowledge with ignorance and death resonates with apocalyptic scenarios. It questions, as do many STS scholars and many artists, the production of vast quantities of data and what it means for our understanding of the world. Increased knowledge does, as has been known since the time of Socrates (Plato, 2014), bring greater awareness of what we do not know, of uncertainties and unknown unknowns. What does this mean for our perception of knowledge, and, in particular, of what we might seek to know through research?

Within epistemology, discussions of knowledge often depart from the justified true belief theory (Gettier, 1963; Klein, 1971). This is described as the traditional definition of knowledge. Knowledge, in this perception, is something that we have good reason to believe is true. However, it deals primarily with cases of propositional knowledge, that is, *knowing that*, as in "S knows that *p*". This theory has less to say about *knowing how*, practical knowledge such as how to ride a bike and other forms of ineffable knowing (Polanyi, 1962; Zembylas, 2014). It deals, fundamentally, with rational aspects of knowledge. Fisher, in *The Vehement Passions*, asks "Could any pair of words seem as natural together as the words 'dispassionate knowledge'?" (2003, p. 1). Emotions are rarely mentioned in public discourse about knowledge; more common are terms like intellect, accumulated data, detached analysis.

This division of emotions from rationality is a typically modern phenomenon, the responsibility for which is often attributed to Cartesian mind–body dualism. As Feyerabend points out in the second of his *Three Dialogues on Knowledge*, "myth, tragedy, the older epics dealt with emotions, facts, structures all at the same time, and they had a profound and beneficial influence on the societies in which they occurred" (1991, p. 113). Feyerabend's interlocutor "B" claims that Stone Age man, in his close relationship to nature, had access to knowledge that we have lost. "The rise of Western rationalism destroyed this unity and replaced it by a more abstract, more isolated, and much more narrow idea of knowledge. Thought and emotion, even thought and nature were separated" (ibid., p. 113). This resulted in a more impoverished language for expression of knowledge, and humanity was separated from nature, Feyerabend argues. When man "after much error" returns to nature, it is "as her conqueror, as her enemy, not as her creature" (ibid., pp. 113–114). This attitude is described by Gadamer (2004, p. 310) as the epistemological stance of "knowledge as dominion". The same modern attitude was identified by Husserl in his 1936 work *The Crisis of European Sciences* (1970) as occasioning, precisely, a situation of crisis. In this tradition, philosophers such as Heidegger, Gadamer and Feyerabend continued critiquing modern science and technology for its fact-based detachment from nature and emotion. Supplements and alternatives to rationalist knowledge views have, indeed, been sought in a number of human and social sciences. Nursing, anthropology and numerous other fields, as well as STS, have foregrounded the vital, and underappreciated, role of embodied, tacit, practical and emotionally engaged forms of knowing.

STS Knowledge Views

The propositional view of knowledge has been challenged since the 1970s by STS scholars, who have focused on the importance of tacit knowledge (Collins, 1974; Vogel & Dennis, 2018) and critical examination of assumptions about the character and state of scientific knowledge (Jasanoff, 2004; Latour and Woolgar, 1986). Tacit knowledge was proposed by Polanyi (1962) as knowledge that is inherently unvoiced and possibly unvoiceable. His classic example was riding a bike, which can be learned through example but has to be practised. Being able to describe what is required to ride a bike is not the same as being able to ride it well. This is true for a lot of the skills involved in scientific work (Polanyi, 1962; see also Collins, 1974; Latour and Woolgar, 1986). Meanwhile, tacit knowledge is perhaps even more of a familiar territory in art and aesthetics (Kjørup, 2006; Schindler, 2015). The practical, embodied, experiential mode of art is typically seen to be ineffable.

Conspicuously concurrent with the focus on different knowledge forms, a distinction between "research" and "science", which clearly favours the former, has been observed by STS scholars. Latour (1998) defines science as "certainty", with characteristics of objectivity, coldness and detachment both from society and politics, whereas research is seen as inherently uncertain, open and creating engagement and controversies. Nowotny, Scott and Gibbons (2001) agree with Latour in considering research to be more open and contested than traditional science and further claim that research is part of society in a way that was unheard of before. They argue that the demarcation between science and society has been increasingly blurred in the last few decades, a tendency intensified by the increased education level worldwide: there are many more individuals out there who are competent in science, so one cannot assume that the "masses" are unknowledgeable and incompetent. This has also led to a greater diversity of perspectives, approaches and transdisciplinary endeavours. "Rhetorically, if not actually, 'science' aspires to unify. But there can be no 'unity of research'; it is diverse and heterogeneous by definition" (Nowotny et al., 2001, p. 68). This leaves the stage open for new knowledge actors, which some argue should include artists.

Artists were, in fact, among the earliest non-scientists to get involved with the natural sciences, through residencies in scientific labs and development of works that use scientific methods. Born and Barry have set the stage for viewing "art-science as a good example of the kinds of practices associated with Nowotny et al.'s mode-2 knowledge production" (2010, p. 104). They consider art-science, involvements of the arts with the (natural and medical) sciences, as one avenue for growing interaction between the institutions of scientific knowledge and the public.[1] As we will see below, bioart projects can be seen as entries in this conversation, through their material use of technologies, interdisciplinary mobilisations and critical engagement with scientific visions.

STS scholar Ezrahi, in "Science and the Political Imagination", presents a hierarchy of modes of knowledge found within our current Western society. He argues that our society has moved "from wisdom through knowledge to information", which "not only tends to diminish the source of knowledge as an agency, but also the agency of the recipient" (2004, p. 257–258). He defines "wisdom" as "a form of knowing or communicating knowledge (I use the term 'knowledge' also as a generic term for all forms of knowing) [that] is characteristically unformalized and even unformalizable" (ibid, p. 255). Ezrahi refers to wisdom as an older, traditional mode of knowing, largely replaced by scientific knowledge during the Enlightenment and one that is more predominant in non-Western societies, such as China. He lists a development, in our time, where knowledge is supplanted by information, the more democratic and accessible but also more superficial, less theoretically attached mode of knowing, and even information is increasingly supplemented by what he calls "outformation". This ungainly term is used to refer to image-led, cinematic, mass-media forms of communication. He considers this to be more closely related to wisdom than the other two modes of knowing, as they are both "rich and frequently intense" (ibid., p. 258) and combine various aspects of experience including the emotional, the cognitive and the aesthetic. However, the stimuli of the outformations are more disconnected and less organized than those of the wise sage. Although Ezrahi mentions that he does not consider the historical shifts between these modes of knowing to be all-encompassing, and indeed, less of a linear process than he sketches it as, he does claim that wisdom, "in the wider sense which may include poetry, art and various esoteric practices" (ibid., p. 265), seems less relevant on the political arena, today, although it still has some space in personal life. While some philosophers have defined wisdom as having extensive rational knowledge and being able to use it well or even as a rational approach to using scientific knowledge for solving the world's problems (Maxwell, 2014), with Ezrahi, and supported by other contemporary philosophers such as François Jullien (2002) and Daniel A. Kaufman (2006), I will consider wisdom as intimately linked to experience and moments of inspiration, and an ability to interpret the significance of various episodes of life, which combines "cognitive, moral, emotional, social, philosophical, and practical references" (Ezrahi, 2004, p. 255).

A potential source of inspiration can be found in Dewey, who, in *Art as Experience*, reacts to the claim that art should be a mode of knowledge but suggests "that in both production and enjoyed perception of works of art, knowledge is transformed; it becomes something more than knowledge because it is merged with non-intellectual elements to form an experience worthwhile as an experience" (2005, p. 302). Wisdom is, precisely, something more than knowledge, based in experience and common sense but also emotional comprehension. Dewey argues that since different artists have different interests, "all aspects and phases of experience are covered" (ibid., p. 197).

Artistic Research: What Mode of Enquiry?

This seems to fit artistic meaning-making quite well. Busch, drawing on Foucault's discussion of properties of art in "Maurice Blanchot: The Thought from Outside" (1990), holds that artists doing research are too easily made to base themselves on science, specifically on "true discourse" (Busch, 2009, p. 5). She argues, instead, for a critical "art or a poetics of knowledge" (ibid., p. 4) that sets out to question the tenets of the academic system. This seems to me to be a fruitful track, but still to some extent leaves open the question of what a poetics of knowledge entails.

While most artists tend to do some research for their projects, the work currently defined as artistic research tends to include larger amounts of research over a greater expanse of time. How different is the motivation for artistic enquiry from that of other academic research? Both the arts and sciences are partly motivated by curiosity-driven research, partly by funding incentives and developments in society, and their products can, sometimes, be indistinguishable but for the context they are presented in. One distinction that has often been made is that, where science has a clear "purpose", in the Kantian inheritance, art is often seen as being "purposeless", "disinterested". The purpose of science is to produce knowledge. And art is often juxtaposed to this, in having a meaning that is "true" at a different level (Heidegger, 2002; Schiller, 1960) and that is considered separate from everyday experience (Dewey, 2004). However, in today's research climate, both the arts and the sciences are also increasingly entwined in society, eschewing their respective traditions of autonomy (Nowotny et al., 2001).

Some simple objections have been made to the idea of artistic research as knowledge production. While many of the methodological steps taken by artists working in labs resemble those of scientists to the point of being impossible to distinguish

(Vaage, 2016), the rigour of artists and scientists take very different forms (Busch, 2009). In scientific research, there is an expectation that researchers will, indeed, come up with propositions. This is seen, in particular, in the focus on written outcomes,[2] especially academic publications, in which the process of research as well as its findings are meticulously laid out.

James Elkins has observed that the "conceptual apparatus of hypothesis and experiment hardly figures in art" (2009, p. 38). Such conceptual apparatuses have, in the past, been seen as core to the scientific method, but, as discussed above, in STS scholarship this is no longer equated with research. The artistic research projects I will discuss below, contrarily, often apply the conceptual apparatus Elkins refers to at an early stage in the process but then expand to concepts at other levels such as societal critique or visions. This process is rarely linear and does not necessarily progress as planned (which can, of course, be said for most scientific research as well).

In the remainder of this paper, I will explore whether we can speak about (certain forms of) artistic research as producing wisdom, through consideration of research processes and output at SymbioticA.

SymbioticA: Open-Ended Processes of Artistic Research

SymbioticA is based within the School of Human Biology at the University of Western Australia, Perth. It has occupied a central position in the field since its establishment in 2000, both due to the seminal work of its permanent artists on the Tissue Culture and Art Project, and its artist-in-residence programme. The artistic research at SymbioticA includes exploring strategies for and implications of presenting biological art in different contexts, and also the further development of scientific protocols and technologies into "artistic tool kits" (SymbioticA, 2011).[3] During my case study I engaged in participant observation, archival research and semi-structured research interviews with the artists, scientists and admin personnel who worked with them.[4]

The SymbioticA website states that it supports "non-utilitarian, curiosity-based and philosophically motivated research" (SymbioticA, 2011). The permanent artists at SymbioticA consider the process of research the core part of their work. To each of the many residents who come to spend a few weeks or months at SymbioticA, they stress that it is a "research and development place, rather than production" (Zurr, interview, 3 May 2013). As academic coordinator Ionat Zurr explained, they prefer for residents "to come with an open-ended project, rather than artists who come and say, 'we want to reach this goal', and get there" (ibid.). Zurr's partner Oron Catts, the director of SymbioticA, described it as causing a bit of a crisis for some residents, who struggled with not having a clear aim in sight. But eventually, he said, most of them reached a point where they realised what "a rich experience it is, that would then inform their practice for many years to come, so I think they reach some stage in their residency where they go through this transformation, actually engaging in this pure, open research" (Catts, interview, 24 Apr. 2013).

Catts, in one of our informal conversations, invoked the Germanic idea of *Wissenschaft*, which is broader than the English "science", and that what they do at SymbioticA might be classified as research but not science. In the ensuing discussion about whether this was really useful language, he said that referring to their work as research is necessary in the academic world in which SymbioticA is based, being affiliated with the university and reliant on applications for funding.

Catts and Zurr have in the course of twenty years explored tissue culturing animal cells in laboratories, resulting in a number of works that stretch the potential of such techniques. Their artworks engage questions around what it means for something to be alive, and how to relate to life at very different scales. In several of their tissue culture works, they included "feeding rituals" and, at the end of the exhibition period, "killing rituals", in order to make tangible the living status of the tiny cells making up their tissue culture sculptures.

Zurr reflected around how "it's very important for us to keep the work ambiguous ... The message is more in the papers that we write, we let ourselves be more, not specific but explicit. But with the work we kind of put a situation there, and that is very open, so it can be read in many ways. I think people realise there was research there" (Zurr, interview). For her, an important part of the aim of her work is raising questions about scientific processes, through reflecting about what they themselves are doing, as well as about what people within the sciences are doing.

What "artistic research" meant to her was immediately connected to PhD projects, especially her own and that of her current PhD student. For her own PhD, she ended up handing in a theoretical thesis, both because she had produced enough text and because it was easier within the system of the University of Western Australia. No clear criteria had yet been made for artistic PhDs, and there were preconceptions among her colleagues "that it's not as rigorous" as scientific research. She did consider it important for artists to develop such programmes, and referred to Chris Salter's endeavours of research creation in Canada as exemplary work "to change this perception that it's not as rigorous as academic research" (Zurr, interview). She viewed it as a very different form of research than that of the biologists they worked with, but, with decades of experience in radically interdisciplinary work, she observed that for every discipline, "you need to be there for long enough to understand about the rigor of that discipline. And also understand ... the bullshit of that discipline" (Zurr, interview).

Scientists' Perspectives

Stuart Hodgetts, the scientific advisor at SymbioticA, pointed out in our interview how scientists are rarely at liberty to play around extensively with a certain idea, as they get their funding specifically to respond to a certain hypothesis and to get results. He explained how in his view, artistic involvement opens up possibilities for him and the other biologists to

> manipulate the system a little bit more there. You know, you're really on a mission to have a finished project, in the same way that you are in science. But the way you get there, or the fact that you don't even necessarily get there, but you found something really interesting along the way – and creative, and productive – that's, that's the real blood of the art coming out. (Hodgetts, interview, 8 May 2013)

Hodgetts considered this important, not just for the artists' process, but also as opening up a space for scientists to work in a different mode. He personally considered his artistic engagements an eye-opener: "Very often we're blind to that creative component to our work" (Hodgetts, interview).

Stuart Bunt, a neurobiologist who was the scientific director of SymbioticA in the first years of the centre and still works at the School of Human Biology, saw two main types of research that the artists-in-residence engaged in: the first being simply to learn new techniques, such as tissue culturing, because they needed it for their planned artworks, and the other being research into the field of biology itself, "how scientists work, how labs work, how people react with animals and animal research, and dealing with living things and nonliving things" (Bunt, interview, 6 May 2013). He stressed how the latter had greatly enhanced the work of some residents, such as performance artist Kira O'Reilly. Simultaneously, the artists challenged scientific assumptions about the good research environment and the rationale behind their protocols. Bunt described how O'Reilly, during her residency, donned a red lab coat and bright red high heels and was photographed inside one of the science labs: "a lot of people got very upset about it, thought it was like, almost like breaking the sanctity of a church, like you wouldn't go in a church and sort of dance around". He thought she had successfully challenged some unspoken rules that people were not even aware of, which revealed "this idea of the laboratory as ... a temple of worship of science" (Bunt, interview).

Artistic Research Practice: Towards a Dog–Human Hybrid

The artists-in-residence at SymbioticA at the time of my case study did not share the permanent artists' perception of artistic research as the very specific activity of academic research with arts outcomes. Benjamin Forster, a young artist who had a rather ambivalent view of academia, argued that "everything we do is research". At art school, he had been introduced to "material practice, which I would definitely equate to research". In his perception of material practice, there is a dialogue between materials and concept; an idea has to be tested in the encounter with materials:

> Even if you have the greatest idea, and you think it'll work really well with cellular biology, and you go in there, and you're like, let's try and make this idea out of this medium, if the medium doesn't adhere, then it doesn't. So ... it's always a dialogue. And I feel like that's research, right? (Forster, interview, 29 Apr. 2013)

Forster's project at SymbioticA was in that spirit of material experiment: he sought to make a hybridoma, a hybrid cell joining human and dog white blood cells.

Forster had done a previous residency at SymbioticA two years earlier, in which he developed the idea for the piece, inspired by the realisation that most artists he knew and admired were rather cynical and looking back to the ancient Greek Cynics, a name derogatively given because they "lived like dogs". According to Forster (interview), this original cynicism, which rejected the confines of society and sought to expose its ridiculous and harmful practices, was something humanity should strive for, and we could benefit from being more like dogs.

The part of the process defined as doing research, thus, engaged with ideas and organisms simultaneously, but it was also, to some extent, a matter of testing whether scientific processes would technically work. This was certainly the case for Forster's project, which involved processes familiar to the biology researchers but with some unusual elements. While fusing cells together to make hybridomas is a relatively old technology, it is mostly used to combine rodent cells with myelomas, human cancer cells. Forster used an artificial myeloma, an immortalised cell line, from the artist Billy Apple, and primary blood cells from a mongrel dog. The dog blood was the less known entity.

During the first residency, Forster had tested most parts of the process. However, some of the protocols he could not quite recall, and others, he was told by the scientists he now worked with, were not the most efficient. He therefore had to repeat quite a few experiments. Forster did not have any scientific background, and relied on guidance from various biologists and technicians. But he did the majority of contact hours with the cells on his own, providing them with new nutrients every few days, checking whether they were looking healthy under the microscope, and pouring some of them into bleach when they became too copious (Figure 5.1).

Some more complex processes tested the skills of Forster's scientific mentors, as well: for instance, when he sought the help of a PhD student to test whether his Billy Apple cells were secreting antibodies, a key function of white blood cells' immunological defence. Being immortalised, they should not, but he wanted to make sure. The dog cells would be secreting antibodies, so if the human cells were not, exposing the hybrid to various antigens could be a quick way to tell whether the dog cells were present in the hybrid. Conceptually, this might not make a difference to his artwork – many bioartworks make claims about scientific processes that may or may not take place (see e.g. Levy, 2006) – but from the perspective of the scientists supporting him, aiming to make the hybrid "viable", if possible, was the obvious choice.

The procedure involved putting the medium the cells grew in into contact with an electrophoresis gel that reacts with protein. This was a complex process that involved several attempts, which also showed both the level of skill and nimble hands needed

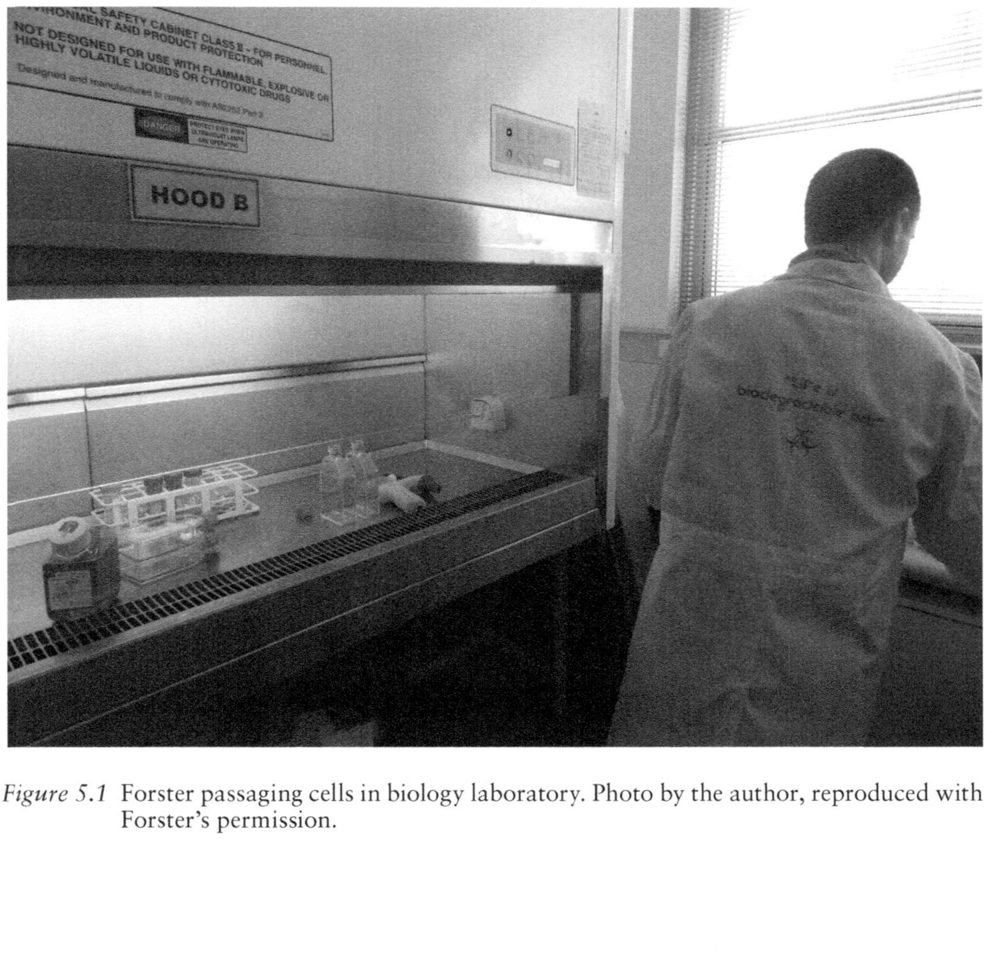

Figure 5.1 Forster passaging cells in biology laboratory. Photo by the author, reproduced with Forster's permission.

to handle such gels and the automation involved in the many machines used during the process. After the results of the first gel were rather shaky, the PhD student decided to repeat the experiment and to make her own gel, which took even longer. In the end, they were able to confirm that the Billy Apple cells did not secrete antibodies. These typical hurdles of biology research, thus, were key to the process of Forster's piece, different though his aims were from those of a natural scientist.

However, the cell fusion that would be the culmination of the project never happened. The white blood cells isolated from the dog blood were few to start with and did not thrive; the artist hypothesised that it was due to the mongrel's old age (Forster, personal communication, 23 Nov. 2015).

The scientific failure of the project had a visibly insignificant, however conceptually profound, effect on the exhibition Forster had been working towards, Semipermeable(+), a SymbioticA retrospective which was part of the ISEA (International Symposium for Electronic Arts) in Sydney in June 2013. For his piece *Kynic*, he had planned to exhibit a single hybrid cell (that would have been invisible to the naked eye), on a slide mounted on a pedestal (Plate 1). Instead, the slide contained a simple message written with a blue marker pen in capital letters:

> 26:05:13 – CELL DEATH
> HARRY CELLS DIED. FUSION FAILED.
> NO IMMORTAL CELLS – PINK DOTS.
> MONTHS OF LABOUR POINTLESS.
> THE RESEARCH IS TO CONTINUE.
> IN SEARCH OF THE KYNIC CELL.

Otherwise, the piece was as the artist had intended, with a crumpled aesthetics of small figurative drawings thrown on the floor, depicting a dog cowering under a philosopher's bust, dogs mating, and the Cynic philosopher Diogenes with a white streak running down the paper next to it (nothing alerted the audience to the fact that this was a trace of the artist's urine). Parts of the drawings could not be seen, as some sheets of paper were completely crumpled into balls.

Both the slide's message and Forster's catalogue text embraced what the SymbioticA artists call "the aesthetics of failure". Neither, however, provided any clear narrative of what the piece had intended to do or what the technical failure consisted of. While leaning over the low fence to make out the crumpled drawings, the audience was thus left to ponder how the merging of dogs and humans might say something about cynicism and about what it means to be human, without much guidance but also without the complications of technical descriptions.

Artistic Research Practice: Bees and Eggs

Nigel Helyer, an Australia-based British artist who was a resident at SymbioticA in 2002, had been loosely connected to the centre ever since. Helyer has a doctor of creative arts degree from the University of Technology Sydney and expressed his belief that practice-based research is eminently possible but often done badly. He attributed this largely to the individualist traditions within the art world, resulting in "an almost complete inability to deal with teamwork. So you find that nearly all of practice-based research in Australia is individuals doing something" (Helyer, interview, 23 Apr. 2013).

With decades of experience as a sculptor and sound artist, Helyer's first SymbioticA work was *Host*, in which an "audience" of several hundred live crickets attended a biologist's lecture on the sex life of insects. At the time of my case study, in 2013, he was working on a few different projects collaborating with UWA scientists. The main one was in collaboration with the Centre for Integrative Bee Research, seeking to have bees create wax structures atop Lego buildings. He had his own three hives of bees in the bee research area at UWA. It was well fenced in, with big yellow warning signs saying "BEES AHEAD!"

> We don the beekeeper suits, hats and all – I am wearing almost all black, and Helyer says the bees don't like black, nor dark hair. He lights up a bee smoker, puffing it about twenty times to make sure it works properly. The precautions add to my sense of apprehension. When we enter the pen where his beehives are, it doubles: there seems to be thousands of bees swarming within those few square meters. However, they are very docile today, possibly due to the heat. After a while, it feels quite natural walking around in there – slowly, because the bees don't like sudden, quick movements. Helyer lifts up the top of each hive, revealing Lego structures glued underneath. They are from the Lego Architecture series, each a model of one of the great buildings of the world. They have been coated with a thin layer of bee's wax, to get the bees started; Helyer wants them to cover the rectangular structures with a more organic mass of bee's wax.
>
> Unfortunately, it's been too hot lately for the bees to produce any honey, which means that they don't build wax, either. In fact, they have "stolen" some of the wax of the Lego structures, instead of building from it. Helyer is disappointed, but there is one partial success: a spiraling structure, like an antelope's horn, descending from the Lego building of the Brandenburger Tor. (Vaage, field notes, 1 Mar. 2013)

Helyer had to wait until the bees started producing honey again with the hope that they would then work further on the wax buildings. The ones we saw were carrying in some pollen at the moment, which Helyer considered a good sign. As it turned out, however, the one wax tower peaking up from the Lego Brandenburger was as good as he would get. This became the focal point of the artwork *Float Like a Butterfly; Sting Like a Bee* (Plate 2). The Lego building and wax structure is visible through a peek hole into a portable beehive. Outside it, a beeswax sculpture of a missile is perched atop a multicoloured Lego ramp, and a butterfly rests peacefully on top of it. A soundtrack of bees buzzing completes the installation. In an evocative article about the piece, published in *Antennae*, a journal for "the non-human turn in the visual arts" (*Antennae*, 2018) which invites texts from artists and scholars alike, Helyer reflects around historical practices of weaponised bees and then swerves into consideration of metaphors of bee colonies in human society, as a starting point for a critique of our "illusion of individuality" (Helyer, 2015, p. 68). He describes his personal experience with the bees, much like I describe mine above, as starting in apprehension, then a sense of immersion within a strange realm. He reports feeling "clumsy and under-evolved" (ibid., p. 69) in the encounter with these creatures "capable of making complex colony wide decisions that leave us baffled" (ibid., p. 70). Few of these reflections, however, are spelled out in the artwork itself. The piece clearly connects insects and warfare, playing on stereotypes of what is harmless and harmful, busy and leisurely. Not least, the title is a direct quote from Mohammad Ali, thus relating to the world of a "civilised" form of fighting, the sport of boxing. The simile of being like a butterfly and a bee evokes ideas about nature that, if we go far into the realm of interpretation, can be seen to provide a different perspective. Rather than perceiving nature as something that man seeks to control, as critiqued by Feyerabend, this piece seems to present it as a source of inspiration, embracing the complex lives of animals in their various aspects, from the vicious to the meditative. Working with the bees, Helyer sought to coax them to perform, to build wax structures on his Lego buildings. When it did not go according to plan, much like Forster, yet less explicitly, he pursued the aesthetics of failure, making lack of control a topic for the piece. But the artwork leaves reflection on these connections up to the audience, and the research process is not described.

The Wisdom of Artistic Research

Artistic research that I consider to be embodying wisdom is not attempting to replicate the unambiguousness, accountability and verifiability that are the often-stressed ideals of scientific research. Rather, it draws on methods of the sciences in order to work across domains, involving both intuitive and rational processes, often embracing messiness, and also challenging the established norms and institutional forms of knowledge. The long-term involvement in the scientific world of some actors, such as Helyer and SymbioticA permanent artists Catts and Zurr, means they can open up a field of encounters and critical discussion with scientists, as well as create artworks through scientific protocols. Part of what makes these practices art is the scope that they take: research processes take place across multiple domains, and the outcome is not fixed by the artists into a single meaning. The unarticulated parts of the artworks are as important and inherent to the piece as what is, or can be, articulated. This goes for interpretations of the artwork as much as the artist's original intentions. Is the buzz of the bees threatening or soothing? What can we read into the honeycomb spire that sears upwards from the Lego Brandenburger Tor? Should we be cynical and, if so, in what sense of the word? Unlike the desired outcomes of most scientific projects, the deliberate ambiguity and multivalency of these artworks means that we can have vastly different experiences and even take diverging ideas from the same piece.

Despite all the more comprehensive perspectives on knowledge and research that have been presented by STS scholars and other academics for decades, it appears that the idea of propositional knowledge as justified true belief is so closely associated with prevailing ideas of research that seeing artistic meaning-making as research is still a stretch for many academics and artists (Elkins, 2009; Garneau, 2008). Considering artistic research as a practice that aims for wisdom might, thus, help create a space for such research that is connected to and complements other academic practices, without having to aim for the same forms of knowledge outcomes.

The inarticulable properties of wisdom that connect it to art: inspiration, experience, the prominence of multisensory and emotional forms of expression but also the idea that this is not something that can necessarily be conveyed directly, that it depends upon inherent qualities of the artist and the receptive horizon of the exhibition visitor, also means that artistic research is not likely to be a strong competitor to academic research as we know it; this is not the aim of most artistic researchers, nor is such research particularly suited for large-scale implementation. Its strength being still in the free, playful mode of exploration, as well as the outcomes' ability to communicate beyond words, it provides a vital supplement in a world where wisdom is a word rarely mentioned, and utility and progress are at the top of most research agendas. Acknowledging how different, yet equally valuable, this form of knowing is to the propositional knowledge we are used to find within academia might assist us in carving out a space for artistic research in the years to come.

Notes

1 I have previously argued that this term, in dichotomously juxtaposing art and science and leaving out other involved fields, is problematic on several counts, see Vaage, 2015.
2 This means most universities cannot allow artists to produce artistic outcomes for PhDs without changing their regulations.
3 For more on SymbioticA and the work of its artists, see Hannah Star Roger's chapter in this volume.
4 My interview technique is inspired by Brinkmann's "epistemic" interviews – a doxastic registration of the interviewee's subjective experiences and opinions. This interview approach seeks to address the respondents as "accountable, responsible citizens" (2007, p. 1117) and is particularly appropriate when interviewing experts in their field. Brinkmann stresses that "an essential normativity runs through the processes and states that we refer to as *knowing*" (ibid., p. 1123; original emphasis). In the encounter with researchers it seems appropriate to acknowledge this normativity and, rather than striving towards the unobtainable goal of not influencing the interviewee, seek to "dig deep" with them in order to arrive at new realizations.

References

Antennae (2018). Antennae: A Decade of Art and the Non-Human: 07–17. Retrieved from http://www.antennae.org.uk/antennae-ten/4594132660 (accessed 15 Feb. 2019).

Borgdorff, H. (2010). Where Are We Today? The State of the Art in Artistic Research. In T. Lind (Ed.), *Forskning och kritik – granskning och recension av konstnärlig forskning.* Årsbok KFoU 2010 [Yearbook for Artistic Research] (pp. 17–31). Stockholm: Vetenskapsrådet.

Born, G., & Barry, A. (2010). Art-Science: From Public Understanding to Public Experiment. *Journal of Cultural Economy, 3*(1), 103–119.

Brinkmann, S. (2007). Could Interviews Be Epistemic? An Alternative to Qualitative Opinion Polling. *Qualitative Inquiry, 13*(8), 1116–1138.

Busch, K. (2009). Artistic Research and the Poetics of Knowledge. *Art and Research: Journal of Ideas, Contexts and Methods, 2*(2), 1–7.

Collins, H. M. (1974). The TEA Set: Tacit Knowledge and Scientific Networks. *Science Studies, 4*(2), 165–185.

Dewey, J. (2005). *Art as Experience* [1934]. New York: Perigee Books.

Eliot, T. S. (1963). *Collected Poems, 1909–1962, centenary ed.* Orlando: Harcourt Brace.

Elkins, J. (2009). Aesthetics and the Two Cultures: Why Art and Science Should Be Allowed to Go Their Separate Ways. In F. Halsall, J. Jansen & T. O'Connor (Eds.), *Rediscovering Aesthetics: Transdisciplinary Voices from Art History, Philosophy, and Art Practice* (pp. 34–50). Stanford, CA: Stanford University Press.

Ezrahi, Y. (2004). Science and the Political Imagination in Contemporary Democracies. In S. Jasanoff (Ed.), *States of Knowledge: The Co-production of Science and Social Order* (pp. 254–273). London and New York: Routledge.

Feyerabend, P. (1991). *Three Dialogues on Knowledge.* Oxford and Cambridge, MA: Blackwell.

Fisher, P. (2003). *The Vehement Passions.* Princeton, NJ, and Oxford: Princeton University Press.

Foucault, M. (1990). Maurice Blanchot: The Thought from Outside. In B. Massumi (trans.), *Foucault/Blanchot: Maurice Blanchot: The Thought from Outside and Michel Foucault as I Imagine Him* (pp. 9–58). New York: Zone Books.

Gadamer, H. G. (2004). *Truth and Method* [1975], trans. J. Weinsheimer & D. G. Marshall, 2nd ed. London and New York: Continuum.

Garneau, D. (2008). Art, Science and Aesthetic Ethics. In S. Caulfield & T. Caulfield (Eds.), *Imagining Science: Art, Science, and Social Change* (pp. 27–29). Alberta: University of Alberta Press.

Gettier, E. L. (1963). Is Justified True Belief Knowledge? *Analysis, 23*(6), 121–123.

Heidegger, M. (2002). The Origin of the Work of Art [1950]. In J. Young & K. Haynes (Eds.), *Off the Beaten Track* (pp. 1–56). Cambridge: Cambridge University Press.

Helyer, N. (2015). Float Like a Butterfly; Sting Like a Bee. *Antennae, the Journal of Nature in Visual Culture, 31*, 65–71.

Husserl, E. (1970). *The Crisis of European Sciences and Transcendental Phenomenology* [1936], trans. D. Carr. Evanston, IL: Northwestern University Press.

Jasanoff, S. (2004). Ordering Knowledge, Ordering Society. In S. Jasanoff (Ed.), *States of Knowledge: The Co-production of Science and Social Order* (pp. 13–45). Abingdon, Oxon, and New York: Routledge.

Jullien, F. (2002). Did Philosophers Have to Become Fixated on Truth? trans. J. Lloyd. *Critical Inquiry, 28*(4), 803–824.

Kaufman, D. A. (2006). Knowledge, Wisdom, and the Philosopher. *Philosophy, 81*(315), 129–151.

Kjørup, S. (2006). *Another Way of Knowing: Baumgarten, Aesthetics, and the Concept of Sensuous Cognition* (Sensuous Knowledge series, Vol. 1). Bergen: Bergen National Academy of the Arts.

Klein, P. D. (1971). A Proposed Definition of Propositional Knowledge. *Journal of Philosophy, 68*(16), 471–482.

Latour, B. (1998). From the World of Science to the World of Research? *Science, 280*(5361), 208–209.

Latour, B., & Woolgar, S. (1986). *Laboratory Life: The Construction of Scientific Facts* [1979], 2nd ed. Princeton, NJ: Princeton University Press.

Levy, E. K. (2006). Art Enters the Biotechnology Debate: Questions of Ethics. In E. A. King & G. Levin (Eds.), *Ethics and the Visual Arts* (pp. 199–216). New York: Allworth Press.

Mäkelä, M., Nimkulrat, N., Dash, D. P., & Nsenga, F.-X. (2011). On Reflecting and Making in Artistic Research. *Journal of Research Practice, 7*(1), Article E1.

Maxwell, N. (2014). *How Universities Can Help Create a Wiser World: The Urgent Need for an Academic Revolution*. Exeter: Imprint Academic.

McNiff, S. (2013). Opportunities and Challenges in Arts-Based Research. In S. McNiff (Ed.), *Art as Research: Opportunities and Challenges* (pp. 3–9). Bristol and Chicago: Intellect.

Nowotny, H., Scott, P., & Gibbons, M. (2001). *Re-Thinking Science. Knowledge and the Public in an Age of Uncertainty*. Cambridge: Polity Press.

Plato. (2014). *Apology*, trans. B. Jowett. Adelaide: eBooks@Adelaide.

Polanyi, M. (1962). *Personal Knowledge: Towards a Post-Critical Philosophy* [1958]. London, Melbourne and Henley: Routledge & Kegan Paul.

Schaffer, J. (2011). Artistic Re-re-research, trans. K. Dooley. In V. Deifel, B. Kraeftner & V. Widrich (Eds.), *An Envelope for Arts, Sciences, Politics and Us: Mixing Realities and Mediating Myths and Methods* (pp. 156–166). Vienna and New York: Springer.

Schiller, F. (1960). *On the Aesthetic Education of Man* [1795]. New York: Frederick Unger Publishing.

Schindler, J. (2015). Expertise and Tacit Knowledge in Artistic and Design Processes: Results of an Ethnographic Study. *Journal of Research Practice, 11*(2), Article M6.

Smith, M. (2008). Introduction. Why "What is Research in the Visual Arts? Obsession, Archive, Encounter". In M. A. Holly & M. Smith (Eds.), *What is Research in the Visual Arts? Obsession, Archive, Encounter* (pp. x–xxvi). Williamstown, MA: Sterling and Francine Clark Art Institute and Yale University Press.

Steyerl, H. (2010). Aesthetics of Resistance? Artistic Research as Discipline and Conflict. *MaHKUzine Journal of Artistic Research, 8*, 31–37.

SymbioticA (2011). SymbioticA Research. Retrieved from http://www.symbiotica.uwa.edu.au/research (accessed 24 Jan. 2019).

Vaage, N. S. (2015). On Cultures and Artscience: Interdisciplinarity and Discourses of "Twos" and "Threes" after Snow's Two Cultures. *Nordic Journal of STS, 3*(1), 3–11.

Vaage, N. S. (2016). Amplifying Ambiguities: Art on the Fringes of Biotechnology (PhD thesis). Bergen: University of Bergen.

Vogel, K. M., & Dennis, M. A. (2018). Tacit Knowledge, Secrecy, and Intelligence Assessments: STS Interventions by Two Participant Observers. *Science, Technology, and Human Values, 43*(5), 834–863.

Zembylas, T. (2014). Forms of Knowing in the Literary Writing Process. In T. Zembylas (Ed.), *Artistic Practices: Social Interactions and Cultural Dynamics* (pp. 112–131). London and New York: Routledge.

6 STS by Material Means

Art Critiquing Science

Hannah Star Rogers

Oron Catts is checking the back and then the front of a white cabinet holding a bioreactor. The cells growing inside were specifically cultured for this moment. The bioreactor should provide an automated system for realizing the conditions and the nutrients necessary for cell life, but the power supply does not seem to be functioning. Catts checks the back again. At last, a tiny light pings on indicating that the bioreactor is finally creating the right conditions for maintaining life. Catts wedges out from behind the cabinet to do final checks and inquiries with the staff about the electrical situation during the night. He is not wearing a lab coat, and we are not in a biological laboratory.

Catts is setting up his joint artwork, produced with Ionat Zurr, "NoArk II" (Plate 3). This artwork highlights a crisis of taxonomy: life in the laboratory defies everyday definitions of the living. The piece contrasts what the artists (collectively known as Tissue Culture and Art; TC&A) call "the semi-living" with dead whole organisms, the sort expected in cabinets of wonder. Behind the red window appears a stuffed crow, a two-headed bird, and a preserved mouse, among others. This is set in opposition to the smaller blue window which holds the apparatus Catts is troubleshooting. The installation is for "Superhuman: Revolution of the Species" organized by the Australian Network for Art & Technology (ANAT) and the Royal Melbourne Institute of Technology (RMIT) at the RMIT Gallery (2009). The exhibit will draw visitors into a labyrinth of collaborations between artists and scientists. These installations were designed to ask questions about the nature of life and the human condition as a living entity in the current time and as future speculation.

Science and technology-engaged artists are practicing STS by material means. These artists share in STS's concerns: who gets to set the agenda of science, participate in its workings, how does science create and maintain its knowledge corpus, and related power structures. They share STS scholars' thematic concerns: corporate and military control of science (®TMark's *Mutual Funds*, 2000; CAE's *The Marching Plague*, 2006), how metaphors and ideas about human ability to intervene shape our understanding of climate change (Catt's *Autotroph*, 2010; Sobecka's *Cloud Machine*, 2012), how gender and surveillance align around pregnant bodies (subRosa's *SmartMom*, 1999), how the role and reuse of model organisms shapes laboratory biology (Zaretsky's *WorkHorse Zoo*, 2002), DNA "fingerprinting" and proprietary enzyme use by crime labs (Vanouse's *Latent Figure Protocol*, 2007–2009), and a plurality of other issues. Many of these artists understand themselves as activists with a specific agenda at the site of a political or ethical question in which science or technology is central. Other artists think of their art as a form of research, or even explicitly as experimentation, about what art can add to public conversations about science. This divide between artist-activists and artist-researchers is reminiscent of conversations in STS wherein applied or public policy-oriented scholars have had overlapping but not identical methods with those who are more theoretically oriented.

These methods of public engagement usually involve reconfiguring social relations to set members of the public in seats of authority in order to stimulate more power-neutral dialogues. Members of the public hear speeches from scientists, policy-makers, and lay persons with special interests. They are given texts to help inform their thinking on given questions. STS scholars write about these issues and the outcomes of such public forums. In short, they encounter arguments in language form from all sides. Why are material means, which seem so effective for persuasion in the world of the artist and the scientist, so often left at the gates of the STS world?

To see these artists who are engaged with science and technology as outside the STS community is a form of boundary-making. Art and science and technology studies (ASTS) is beginning to unpack some of this work and its consequences for STS. One easy to spot difference between STS scholarship and the work of artists is the use of materials to make STS arguments. Artists' materials go beyond the recorders for interviews or laptops or office supplies that feature so prominently in the STS collective research proposals. They are like scientists' immutable mobiles (Latour, 1986), a vital feature of the production and dissemination of knowledge. The very stuff that STS's erstwhile informants, scientists, engineers, industrialists, designers, and users have aligned and synthesized to create new knowledge is left out of STS methods. Materials, or as Calvert and Schyfter have called them, "tangible artifacts" (2016), are the ways that science is made and the basis for the technologies which STS scholars have pointed out to make arguments.

Art and Science and Technology Studies: Work with Materials

Not all STS scholars have been persuaded that their only place was in the world of texts. The turn toward making was the focus of the inaugural "Making and Doing" sessions and displays which were added at the Society for the Social Studies of Science (4S) conference in Denver (2015). These interests have proved persistent and are now part of each annual 4S conference. The stated aim of "Making and Doing" was to emphasize "scholarly practices for producing and expressing STS knowledge and expertise that extend beyond the academic paper or book." The variety of material engagements on display at 4S ranged from films and games to exhibits and literary works. This exposed the extensive range of materials being put to use in the STS community's projects.

Yet the majority of STS scholarship appears only on the page. STS considers materials in an abstract sense: STS scholars often do not explicitly use materials as epistemological tools. Surely no one would argue that ideas conveyed through linguistic means are less potent than materials but neither would STS scholars expect their rhetorical tools to displace the potential of materials. Traditional informants whose work is through their materials may well have much in common with artists who make arguments through their artworks. In both cases, the materials they manipulate are often supported rhetorically with the written or spoken word. In this sense, artists fit neatly into STS studies on the interplay between artifacts and practice.

STS has a history of integrating new subjects for analysis and adding new methods from a variety of disciplines (history, sociology, anthropology, philosophy, and visual culture). Most research involves a combination of methods, and the blind spots of each method have been critiqued in turn. Engagement with material practice and its potential benefits in reaching the public and in making arguments in STS communities will no doubt also contain pitfalls worth analyzing. The participant-observer method with its emphasis on being enrolled in the practices of informants (DeWalt & DeWalt, 1998, pp. 109–136) risks the analyst being folded into the scientific narrative. In STS, this method often involves direct contact with laboratory methods and technical operations. The participant-observer method implicates researchers in methods that require materials. Its usefulness in analysis is direct, but it also extends the social benefits of shared experience. Artists' engagement with materials and the power that these engagements have to reach the public also risks the possibility of complete identification with the informant's perspective. In the case of bioartists, this involves the tension between using and critiquing scientific materials. It is these materials, or at least the suggestion of them, that is crucial to attracting the public.

It is this point about public engagement which has received the most attention in terms of the potential of materials to attract, enliven, and embody debates. But it is important not to leave aside the power material engagements may have to make arguments inside the STS field. Just as participant-observer work is an accepted tool, working with materials could become a fundamental practice for STS. More needs to be understood about its potential to create a significant grounding for how it might work as an STS method. Throwing the tent open to contemporary artists who engage with science and technology gives us an opportunity to consider the potential value of material methods in STS.

The isolation of STS scholars from communities of critical science-engaged artists is beginning to change as more STS scholars become focused on the potential for dialogue sparked by organizing the public around artworks. For some, this is simply a matter of awareness. So much artwork with a critical edge exists across science disciplines that many STS scholars could immediately benefit from exchanges. There are artists working in a range of areas from physics and chemistry to urban planning and energy, so STS scholars need only be on the lookout for such artworks in their areas of interest. More formal conduits for exchange of ideas could also be initiated. Some are already in the works like the conference sessions mentioned above, but much more can be done. Examples like the more intensive collaboration structures of *Synthetic Aesthetics* (2010–2014) and *Emerge: Artists and Scientists Redesign the Future* (2012–present) are projects facilitated by STS scholars to observe and intervene in the interactions between art/designer and scientist collaborations. They have the potential to expand these conversations into a triangle of exchanges between artists, scientists, and the public.

Political conditions around the charge of STS scholars to work with the public also favors this work at this moment. The rise of public forums on science and technology issues in the US context and the Responsible Research and Innovation (RRI) framework in Europe have raised the possibility for new institutionalized ways to reach the public that are amenable to artistic contributions. Some limitations of these methods as they stand have been explored by Brown (2015), Mouffe (2013), Durant (2011), and Lövbrand & Beck (2011).

One new way that STS might attempt to engage the public is though conversations based around objects created by artists and designers who know this area well. Cultural probes, some of which use design and art methods, have been deployed with members of the public (Halpern, Erickson, Forlano, & Gay, 2013). They have yielded good results in opening conversations with those less familiar with particular scientific debates. The 2018 National Academies report revealed a rich system of art and science collaborations underway in higher education settings. The report encourages further capacity-building and attention to these modes of teaching and learning. Observation of those capacities being developed among members of the public could be both an object of study and inspiration for methods in STS.

Yet these possibilities for integrating STS and art also have the potential to instrumentalize art, to turn artists into communicators or even advertisers. Western capital-intensive science has a history of enrolling historians and sociologists to celebrate its work in hopes of bringing public opinion along. Its treatment of artists is no different. The disciplining of the artist (Smith, 2004) or artists who have been asked to glorify the achievements of science make quite a different contribution than those whose work harnesses the power of social, political, and ethical critique to material practice (Da Costa and Phillips, 2008).

STS community interactions with artists have thus far been primarily at the individual level as informants (Rogers, 2011a; 2011b; De Ridder-Vignone, 2012; De Ridder-Vignone & Lynch, 2012; Halpern, 2012). This scholarship could be expanded if artists were seen as a category of informant working towards some of the goals of the STS community. Artist-informants have a great deal to add to understandings of science, but their contributions could also be seen through a lens of recognition of how much STS scholars and artists have in common. These artists have already been the object of STS study from philosophical (Hannah, 2013; Vaage, 2016), sociological (Rogers, 2012), communication studies (Halpern, 2012), and geography (Dixon, 2009) perspectives. What they have not been so far, however, are partners in a shared enterprise.

When STS scholars cooperate with artists, often around public policy and curatorial projects, they have tended to focus on locating individuals who could add to STS projects. In order to gain access to a more complete understanding of the ways artists and art communities might work with STS, STS scholars must also begin to engage them at the institutional level. These engagements must reckon with the considerable variety of these artists' material practices, from digital art to artist-created citizen-science street labs. A focus on medium can yield many new insights. But STS scholars should not neglect the often-critical content of those works. This is after all the main focus for the actors in question. Even as these artists work to obtain new skills and use new materials, their *raison d'être* is the creation of new critiques.

Where might STS scholars start in reaching out to artists for more interactions around issues of collective concern? These artists developed their own standards of work, ways of settling priority/originality disputes, education/credentialing, and a set of technical skills with accompanying shared theoretical understandings. They have created the standards that STS scholars use to identify a subgroup of scientists and are, therefore, an excellent place to start examining art using STS tools. These artists have knowledge of the scientific jargon and laboratory skills which allow them to create artwork using mediums like complex data sets, nano-imaging, tissue culture, digital interventions, and synthetic biology. They are technical experts, not only in their exchanges with the public but also, to varying degrees, when they interact with scientists. This interactional expertise (Collins and Evans, 2002) is similar to the skills Steve Epstein's activists acquired in order to enter scientific debates around AIDS (1996). This interactional expertise creates further opportunities for lab access, equipment borrowing, and an exchange of ideas.

STS and Bioart

This chapter considers a particular group of contemporary artists, known as bioartists, in order to examine the specific possibilities for overlap between STS scholarship and artists who are engaged with science and technology. Bioartists is a descriptor offered by artist Eduardo Kac in his 1997 *Time Capsule* and repurposed as a critic's term for a loosely connected group of artists who work with the ideas and materials from biology with a focus on the critique of biotechnology. In particular, I have derived much of the experience that makes up this case from my ethnographic study of the artists, scientists, humanists, and administrators at SymbioticA, a wet research lab at the University of Western Australia (UWA). Lab studies are an excellent way to dive into the world of these artists.

SymbioticA brings together artists and scientists for collaborative wet lab work. SymbioticA's website describes the organization and space as "an artistic laboratory dedicated to the research, learning, critique and hands-on engagement with the life sciences."[1] This established bioart community is dedicated to exploring biological technologies and scientific knowledge from an artistic perspective. SymbioticA is a collection of experts and equipment with stable access to institutional gateways, such as ethics review boards and funding. A myriad of people circulate through the facility, bringing their own ideas and techniques. They participate in individual research and structured activities, such as weekly meetings and laboratory protocol training.

The laboratory is located in the School of Human Anatomy and Biology at UWA. As the first artistic research laboratory of its kind, it has served as a model for other ventures, including the Coalesce Center for Biological Art at New York State University at Buffalo (UB) and the Bioart Initiative at Rensselaer Polytechnic Institute (RPI). Both of which were founded by former SymbioticA residents. Several art-science collaborations, including the Swiss Institute Artists in Labs (AIL) in Zurich and Le Laboratoire in Paris have emerged in the last decade, but SymbioticA is a unique resource for artists as the first functioning wet biology lab at a research university.

SymbioticA has both dedicated lab space and several shared lab spaces, including the animal research facility and the spectroscopy laboratory. The practice of sharing space is typical of all the research groups at the School of Anatomy and Human Biology. SymbioticA was formally founded in 2000 by cell biologist Miranda Grounds, neurologist Stuart Bunt, and artist Oron Catts. SymbioticA has grown in scope and size to encompass research activities, graduate and undergraduate courses, supporting visiting artists (known as "residents," a common art world parlance for institutional visitors), sponsoring arts programs, and hosting meetings between scientists and artists.

SymbioticA offers both material resources and a skills community to the people who are involved in their network. It is well recognized in STS scholarship that life science laboratory skills contain considerable tacit knowledge (Vogel, 2008). So, it is no accident that it has been important for artists who wished to work with tissue culture, genetics, and other lab techniques to have access to other persons with the requisite skill sets. SymbioticA offers these practitioners the lab equipment itself, which is often a major cost and regulatory hurdle for individual artists even in the university context. Perhaps more importantly for visiting artists, SymbioticA offers the cooperation of people, including lab managers, technicians, experienced scientists, and artists who have worked with specific protocols. It is the hallmark of much bioart that it is created with non-standard or DIY lab equipment but obtaining the skills to do this can be nearly impossible without the important training received from other practitioners. Bioart progenitors Zurr and Catts were educated in Charles Vacanti's lab at Harvard University, which became famous through the popular news for the "Vacanti mouse," a tissue-culture ear implanted on a mouse's back which many in the public mistook for genetic engineering.

Because the artists currently produce artworks at SymbioticA, complications around transporting materials often arise. Owing to Australia's strict regulations on the movement of living materials (both in and out of the country as well as across territorial lines), the living components of their artwork must often be created at the site of the display. This is sometimes done through local universities and other research institutions or by purchasing local materials that are either easily put together or grown in makeshift laboratories inside the museum. In either case, technical skill is required.

The politics of these artists involve advocacy of the democratization of science. They are committed to DIY tissue culture and, in particular, to attempting to transfer those skills to other artists and the public through workshops. The reason for this belief probably stems from the artists' early experience with at-home experimentation. Trends toward DIY, particularly DIY synthetic biology and its relationship to amateur work, initially raised the hackles of many conservatives and the interest of the FBI and the Homeland Security Service (Pollack, 2010) but was eventually approved by the Obama administration (Saletan, 2011). The implications of these artists' work in dealing with institutions has the potential to teach STS scholars a great deal about the power of materials to garner the attention of the public and the state apparatus, as well as offering at least rhetorical models for citizen engagements with biology.

But however positive the politics of democratizing laboratory practices may be for bioartists, there are practical problems with the artists' claim to DIY skilling. The artists' major projects were not created in basement laboratories. Even if they were created in street labs or community facilities, many of the projects are possible because of the artist's previous lab experience. Indeed, perhaps a majority of the artworks require sophisticated materials. If they can be executed in non-lab settings, they are certainly facilitated by scientific know-how and sterile conditions. Artists' claim to DIY skills that make labs unimportant and set them up as simply structures of power in which the public is locked out of the proverbial lab paid for by taxpayer dollars is a stretch. It is the equivalent of the lone-genius scientist or artist insisting that their work requires no community or shared knowledge. It represents self-fashioning more than a reality of work practice. Laboratories like SymbioticA and equipment, protocols, and troubleshooting ideas obtained from scientists are vital to these artists' work.

This should come as no surprise to STS scholars since the importance of habitual apparatus use and transferable lab protocols has been a research focus since Harry Collin's investigation of the trouble of replicating the TEA-laser (1992). Most SymbioticA artists had excellent training and resources before they ever showed their first work, and they have been able to pass on, both tacitly and through written protocols, the knowledge and skills it takes to have what Zurr explained to the bioarts class which I attended as part of my ethnography of SymbioticA as "a feeling for tissue culture". This turn of phrase echoes Evelyn Fox Keller's biography of Barbara McClintock (1984) so closely that it would be hard to maintain that the two were unrelated, even if this language is not a direct homage.

STS scholars who are accustomed to scientists judging each other's outputs on whether they "work" in a technical sense are sometimes confused by what artists in this area consider a finished work. Though STS scholars are used to conceptual and theoretical ideas as publication-ready in the science world, they frequently hold artists to different standards. Some artists in this space work entirely or primarily conceptually with some evidence through prototypes and proof of concept. Others are insistent that every component must function in order to consider the piece complete. Understanding that what constitutes "working" is partly the result of an artists' identification with a lineage of artmaking, much as a scientist might identify herself as a theorist or an experimentalist, is important to STS scholars' interpretations and potential conversations with artists.

Artists often share concerns about the way that science is practiced and technologies are employed that overlap with concerns STS scholars have independently raised. For example, SymbioticA artists' work have developed certain attitudes towards ready made science equipment that were a departure from the attitudes of the scientists I interviewed who worked in the same labs. During my initial tour of one of SymbioticA's lab spaces, Catts expressed his disdain toward what he terms "white goods," by which he meant commercially produced biotechnology appliances that separate the user from the process. For example, many bioreactors and imaging equipment give the impression of working with a computer in terms of styling and the way that cultures are loaded into the apparatus. Catts complained that scientists and technicians do not need to know how the equipment works. His recognition of the metaphors implicit in the designs of some scientific equipment may flow from his training as a designer and his attention to the other resonances besides function that an object may offer a user.

Through the course of my ethnography, I repeatedly encountered Catts's concerns about these issues around other lab appliances. The artist complained about the ideas the slick red convertible-like design of lab equipment invoked as he pointed out the highly stylized machines that processed samples in UWA labs. He particularly made fun of the corporate tiger logo that embellished the Melbourne lab incinerator. Most interesting was Catts' critique of biology equipment that gave the user the sense that working with living things could be clean and easy. He described bioreactors that reminded him of the smooth lines of an Apple product CPU. Specimens could be loaded like CDs and left for an appointed length of time. This just made life look too easy. Catts connected his discontent with the lab tools design with the metaphor of the genetic code: these computer-like biological machines were depicted as able to process life forms into data through a strange metaphoric mapping of life as code.

But for him, what STS scholars would call "black-boxing" (Latour, 1999) is problematic since it creates distance from the life that is being manipulated in the lab. Catts and the other artists used this equipment but they had a special awareness of the work these instruments did beyond their advertised uses. These artists also had a special interest in repurposing and reusing older equipment. For example, they were interested in creating a bioreactor from basic components as opposed to purchasing one from a commercial outfit. This was in part because they felt it could be done more cheaply and with results which would allow them to more directly experience the apparatus operations by offering viewers a look inside.

One exemplary case was that of the artwork *Victimless Leather* (2004). For this piece, the primary glass vessel in which a tiny rat-skin jacket was grown was salvaged from UWA scientist Arunasalam Dharmarajan's apparatus for perfusing rabbit ovaries. Perhaps this scavenging behavior by the artists had begun out of thrift but it developed into an interest in creating more impressively complex items through repurposing. It also meant that at the time of my series of interview with him, Catts was considering creating a toolkit in an attempt to create more access for other artists and amateurs to tissue culture as a medium by avoiding some expensive lab tools.

Bioart at SymbioticA: The *Pig Wings Project*

For SymbioticA artists, two parallel tensions exist: between using and critiquing science and between serving and critiquing science. The Wellcome Trust rejected the tissue-culture artwork *Pig Wings Project* (2000) from an art and genetics show for the foundation's gallery. The bioart group Tissue Culture and Art had been solicited to participate in a Wellcome Trust exhibit despite specifically being tissue-culture artists and, therefore, not working with genetics. Upon reviewing their project, which was a critique of the hype around genetics, the artists' work was rejected from the show on aesthetic grounds. This led to a legal dispute which has made many of the related documents unavailable. In this episode, there was a controversy over who could represent science, in the form of the human genome project. SymbioticA artists believed that their role as artists involved a right or even a duty to critique the powerful, in this case science and the Wellcome Trust's involvement in it.

While it is unusual that the artists' work was not included after they had been invited to submit, it is entirely possible that the curators had other ideas about why the artwork should not be included. Because a lawsuit makes it impossible to review the complete set of documents surrounding the case, access is available only to the documents the artists put forward (Zurr & Catts, 2005). According to TC&A, the artists the Wellcome Trust chose for the gallery were included in the exhibit if they reproduced what the Wellcome Trust believed to be the public opinion of the genome project. Put more coarsely, as the artists saw it, the Wellcome Trust gallery wanted artists to advertise the importance of the human genome project, whether they saw it as positive or negative. The Wellcome Trust's encouragement of work on the subject was contingent on the artists' role in supporting the idea that the genetic age was an important turn in human history. Implicitly the presence of artists would show that genetic discoveries were worthy of artistic reactions and artworks, be they celebratory or cautionary. The Wellcome Trust gallery did not imagine a position that engaged the genome project as anything except important. They believed that they understood genome science and even what the public would potentially think of it. Their selection of works ranged from laudatory to critical of the private genome interests' intentions to patent genes, as opposed to the Wellcome Trust's supposedly more benevolent public science investments.

A number of art institutions supported more critical genetics-related artwork during this time which did ask questions about the real consequences or lack thereof. Among the most prominent was *Paradise Now* (2000), which brought 39 artists to

Exit Art gallery in New York just a few months after the completion of the human genome project. The exhibit received considerable media attention and traveled to a number of locations in the US and Europe. Contributing artist Natalie Jeremijenko raised the question of what artists have that positions them to contribute to such debates: "What is it that the artists have that these corporate interests are interested in? It is not the art; it is the access to the public imagination" (Jeremijenko, 2000). Jeremijenko's suggestion is also applicable to scientists and science institutions. They will continue to pursue artists for their role as perceived conveyers of relevant scientific information and creators of attitudes and imagined possibilities for science and technology. In this model, these actors always have a duty in their role as artists to produce work critical of power rather than creating work that promotes scientific interests. For bioartists, that power has been identified as science. In this case, the artist is at once under siege by powerful institutions to produce work that they perceive to be in the service of either public information or engagement.

Accordingly, the *Pig Wings Project* was intended by the artists to defy expectations: the audience's expectations of pigs with wings was met with tiny cultured tissues, and the Wellcome Trust's expectations of a pro-genome project piece met with a piece that questioned how artists are used for science-related public-relations. The story of the *Pig Wings Project* demonstrates two views of artists in relation to science. Both SymbioticA artists and the Wellcome Trust gallery expressed beliefs about the role of the artist in contemporary science issues (Zurr & Catts, 2005). The artists report that they find the Wellcome Trust's contention that the artists should represent the public's beliefs about science strange. This is likely because of the artists' nearly opposite belief that the role of artists in biological materials is to critique science and the claims science makes for society. For the Wellcome Trust, art can be put to use in the service of science.

The presence of art celebrating or critiquing the project served as a layer of what the artists called "genohype." Artists in the service of science is not at all the conception of SymbioticA artists, who are used to a world in which they collaborate with scientists, have access to scientific resources, and even hire scientists to work on their art projects or teach in their workshops. Though there are many cases of artists who work directly for scientists or with the intention of documenting and creating aesthetic celebrations of science, many more have a more complex view of their relationship to the scientific subject. STS scholars might take note of the ways these relations mirror the divide between the chroniclers of science and those whose critiques directly engage scientific ideas and contexts.

At SymbioticA, scientists and technicians often work for artists, but there are also cases of collaboration in which artists and scientists work together in collaborative partnerships. The latter was the case in the project *Fish and Chips* (2001) and its successive iterations known as *MEART* (2004–2005). In this work, fish neurons control a large drawing arm that makes marks which were posed as fish-robot created artworks. This work brings together ideas from neurobiology and robotics to question the nature of creativity and the nature of minds. In interviews, the collaborators recalled the work and resulting credit being shared equitably. Both conceptual input and technical know-how were contributed by people who identified both as artists and as scientists. These equitable contributions and credit sharing helped set SymbioticA as a collaborative space in motion, but many other models of art–science collaboration have produced outcomes at SymbioticA.

The model of scientists and technicians working for an artist was the case when the leading body-artist Orlan came to SymbioticA to create her *Harlequin Coat* (2007), prototype of a biotechnological coat, consisting of *in vitro* skin cells grown in colored diamond shaped petri dishes. The work was designed to inquire about the biology of skin as it is implicated in questions of race and culture. According to accounts by informants familiar with the eminent artist's visit, she viewed her work as primarily conceptual. As Orlan put it on her SymbioticA project website: "Can skins of different colors be cultivated? What kind of information can be obtained from the donors? Can a person still be the owner of his or her cells? Does self-ownership continue to exist at the fragmented level? How are such issues perceived in various countries, and especially in the context of a non-western viewpoint?"[2] The philosophical dimensions of her work are provocative, and, for some observers, her technical methods were as well.

Orlan commissioned elements of her artworks from SymbioticA-affiliated scientists and technicians. This process is quite common in other areas of contemporary art where pieces may be executed by a firm, workers, or other artists employed by the named artist. However, the informants I interviewed in connection with Orlan's residency had varying degrees of respect for this type of process in this context. Some interviewees contended that it was not possible to be sufficiently "implicated" in the matters of life and death of laboratory specimens and cell cultures involved with this type of work without hands-on experience. To an STS scholar, this does not appear dissimilar from a scientist who conceives of a new experimental design and hires others—technicians, for example—to execute the design. The question of how much technical know-how is required to create science-engaged art is similar to the perennial question in the STS field of how much technical knowledge is necessary to create good STS scholarship. Even the internal challenges of the worlds of these artists and STS scholars have a great deal in common.

Conclusion

The activities of artists with physical materials, as opposed to exclusively rhetorical methods, may well hold insights relevant for STS analysis but also the potential for new methods. The power of materials to reach and convince others is not something STS scholars ought to leave aside. As art becomes a new subject for STS analysis by way of ASTS, an opportunity opens to consider the role artistic methods might play in STS work.

To recognize the networks, theories, and bodies of work that artists have created in the face of many of the issues of concern in STS is to open the circle of STS wider. Avoiding considering artists' work as an augmentation or footnote to our collective scholarship benefits everyone. It avoids policing the boundaries of the field in ways that diminish STS's ability to learn from their contributions and to reach the public. Like scientists, SymbioticA artists gather material and rhetorical resources to make their projects count as critiques of science, while employing scientific methods, and yet they position their work as artistic practice. SymbioticA's goals of drawing attention to the way biology works with living things, focusing attention on biotechnology's corporate control, and diagnosing utopian views of the possibilities for science are made possible through the use of scientific materials and practices, which both bolster their position as knowledgeable practitioners and attract attention to their political agendas.

The artists at SymbioticA have created a fused identity, as opposed to seeing themselves as working in multiple roles. This flexibility allows them to both move biological practices into art galleries and make political points about science using artistic logics. Many of these artists trained in both art/design programs and in tissue-culture laboratories before having access to their own lab space at SymbioticA. STS scholars are in an excellent position to theorize about bioartworks and the works of other contemporary artists who engage with science and technology. These artists have a dual lineage in art and in science, both of which need to be treated in interpreting these works. As art historians tend to lack training in the history of science and technology, STS scholars can provide insights about these artworks.

Even some of the complications faced by artists are mirrored by the situation of STS scholars. Both are often beholden to experts, labs, or institutions for research access, even as they attempt to stand outside those relationships and institutions and provide critiques of their inner workings. No less than STS's attempt to show how science functions, these artists are attempting to use their practices to engage deeply with the form and content of scientific work. The categories of art and science are used by SymbioticA bioartists to create the conditions for critiquing biology through material means. Works like those at SymbioticA should be considered STS by other means: that is, these works engage some of the same issues that science studies engages but do so not by publishing papers but by vesting their ideas into physical and tactical objects. Taking artworks into the network of STS offers many advantages in reaching the public through material engagements and in deepening STS's faculties through new perspectives. STS has grown through inclusion. Critical instantiations of contemporary artwork deserve STS's attention. Interactions with their creators have the potential to broaden the community of STS by adding practitioners with STS interests who have new sets of skills in working with materials. This process of analyzing STS work by other means has the potential to extend our scholarship into new areas and enrich our practices with new methods.

Notes

1 www.symbiotica.uwa.edu.au.
2 www.symbiotica.uwa.edu.au/residents/orlan.

References

Brown, M. (2015). Politicizing science: Conceptions of politics in science and technology studies. *Social Studies of Science* 45(1), 3–30.

Calvert, J., & Schyfter, P. (2016). What can science and technology studies learn from art and design? Reflections on "Synthetic Aesthetics." *Social Studies of Science*, 47(2), 195–215.

Collins, H. (1992). *Changing order: Replication and induction in scientific practice*. Chicago: University of Chicago Press.

Collins, H., & Evans, R. (2002). The third wave of science studies: Studies of expertise and experience. *Social Studies of Science*, 32(2), 235–296.

Da Costa, B., & Phillips, K. (2008). *Tactical biopolitics: Art, activism, and technoscience*. Cambridge, MA: MIT Press.

De Ridder-Vignone, K. (2012). The images and art of nanotechnologies. *Leonardo*, 45(5), 431–432.

De Ridder-Vignone, K., & Lynch, M. (2012). Images and imaginations: An Exploration of nanotechnology image galleries. *Leonardo*, 45(5), 447–454.

DeWalt, K. M., & DeWalt, B. R. (2011). *Participant Observation*. Lanham, MD: Rowman & Littlefield.

Dixon, D. (2009). Creating the semi-living: On politics, aesthetics and the more-than-human. *Transactions of the Institute of British Geographers, 34*(4), 411–425.

Durant, D. (2011). Models of democracy in social studies of science. *Social Studies of Science, 41*(5), 691–714.

Epstein, S. (1996). *Impure science*. Berkeley: University of California Press.

Halpern, M. K. (2012). Across the great divide: Boundaries and boundary objects in art and science. *Public Understanding of Science, 21*(8), 922–937.

Halpern, M. K., Erickson, I., Forlano, L., & Gay, G. (2013). Designing collaboration: Comparing cases exploring cultural probes as boundary-negotiating objects. In *CSCW '13, 2013 Conference Proceedings* (pp. 1093–1102). San Antonio, TX: ACM.

Hannah, D. (2013). Performative Experiments: Contemporary Art and the Aesthetics of Scientific Experimentation (unpublished PhD dissertation). Columbia University.

Jeremijenko, N. (2000). A Response to the Paradise Now Exhibition. *Paradise Now/Invest Now*, 24 Oct.

Keller, E. (1984). *A Feeling for the organism: The life and work of Barbara McClintock*, 10th anniversary ed. New York: Times Books.

Latour, B. (1986). Visualization and cognition: Thinking with eyes and hands. *Knowledge and Society, 6*(6), 1–40.

Latour, B. (1999). *Pandora's Hope: Essays on the Reality of Science Studies*. Cambridge, MA: Harvard University Press.

Lövbrand, P., and Beck, S. (2011). A democracy paradox in studies of science and technology. *Science, Technology and Human Values, 36*(4), 474–496.

Mouffe, C. (2013). *Agonistics: Thinking the world politically*. London: Verso.

National Academies of Sciences, Engineering, and Medicine (2018). *The integration of the humanities and arts with sciences, engineering, and medicine in higher education: Branches from the same tree*. Washington DC: National Academies Press.

Pollack, A. (2010). U.S. Bioethics Commission Gives Green Light to Synthetic Biology. *The New York Times*, 16 Dec.

Rogers, H. (2011a). Amateur knowledge: Public art and citizen science. *Configurations, 19*(1), 101–115.

Rogers, H. (2011b.) Art or science? The practices of tactical media. In L. M. Dolling (Ed.), *Science, technology and the humanities: A new synthesis*. Hoboken, NJ: Stevens Institute of Technology.

Rogers, H. (2012). The Practices of Art and Science (unpublished doctoral dissertation). Cornell University.

Saletan, W. (2011). Faking Organisms. *Slate*. Available at http://www.slate.com/articles/technology/future_tense/2011/02/faking_organisms.html.

Smith, P. (2004). *The Body of the Artisan: Art and Experience in the Scientific Revolution*. Chicago: University of Chicago Press.

Vogel, K. (2008). Iraqi Winnebagos of death: Imagined and realized futures of US bioweapons threat assessment. *Science and Public Policy, 35*(8), 561–573.

Vaage, N. (2017). Fringe biotechnology. *BioSocieties, 12(*1), 109–131.

Zurr, I., & O. Catts. (2005.) Big pigs, small wings: On genohype and artistic autonomy. *Culture Machine*, 7. Available at https://culturemachine.net/biopolitics/big-pigs-small-wings.

Part II

Practices

7 Material Systems

Kinetic Sound Art and STS

Jon Pigott

Introduction

Kinetic sound sculpture is a mode of practice not easy to categorise or identify clearly. This is partly because, like many other sound-art related practices and even the overarching genre description of 'sound art' itself, it is a fusion of traditions, materials and activities which are not easily tied to a specific time period or group of artists (Licht, 2009, p. 3). The field can be seen as benefitting from a range of varied influences from music, art and technology. These include examples of eighteenth- and nineteenth-century musical automata, early twentieth-century Dadaist and Futurist projects such as Luigi Russolo's *Intonarumori* (*c.* 1913) or Man Ray's *Indestructible Object* (1923), custom and sculptural musical instruments such as those made by Harry Parch or the Bashet brothers as well as experimental composition and film making. A further difficulty in identifying the field is that by the time kinetic sound works had become an identifiable mode of practice by a range of practitioners in the mid to late 1960s, critical opinion had begun to favour more conceptual, textural and systems-driven approaches within the arts. This left the world of sounding, moving, object-based sculpture looking decidedly old fashioned, and as a result it has been argued that the field struggled to achieve proper critical representation (see Chau, 2014; Keylin, 2015; Cox 2013; Pigott 2017).

Despite this, artists and composers have remained intrigued and inspired by the creative possibilities emerging from moving, sounding assemblages of objects and technologies and the energetic translations that occur within such systems. Some canonical examples from the 1960s are described below, but it is also worth noting the rich breadth of contemporary activity from practitioners including Max Eastley, Godfried-Willem Raes, Pierre Bastien, Gordon Monahan, Trimpin, Peter Bosch and Simone Simons and Zimoun among others which contributes to and maintains a mode of practice identifiable as kinetic sound art. Of course, as with all genre descriptions of creative practice, hard delineations of what counts as kinetic sound art are neither helpful nor necessary, and the field continues to exist within and alongside a rich context of technologically astute media-art practices all of which despatch electronics, sound, objects and performance as and when they see fit.

This paper will explore how sensibilities and approaches from science and technology studies (STS) can help to understand and identify the field of kinetic sound art, extracting specific concerns from a broader milieu of 'dematerialised' and 'expanded' art practice of the late 1960s and 1970s (see Lippard & Chandler, 1967; Krauss, 1979). This is approached by developing the idea of the 'material system' identified in the work of STS scholars Bruno Latour and John Law and relating it to the object- and material-based technological systems of kinetic sound art. Also, kinetic sound art will be shown as a possible way of extending existing STS work around technologically engaged creative practices such as circuit bending (see Pinch, 2016; Parikka & Hertz, 2012). This unapologetically non-expert electronic intervention into battery-powered toy musical instruments and games to explore new sounds has been well examined by STS. The kinetic sound art described here occupies some similar territory to circuit bending (the re-appropriation of existing technologies, for example) but also represents part of a wider field of artists working with engineering and constructing technology. Exploring this lineage of technologically engaged art practice offers further opportunity for an STS of the arts. A first-hand case study of an original kinetic sound piece by the author titled *Electromagnetic Interrogations* (2011–2014) will allow for further reflection on this kind of artistic construction of technology as well as a consideration of how the making of a technological artwork and the exploration of related STS influenced ideas can be thought of as a single 'method assemblage' (Law, 2004, p. 13).

Kinetic Sound Sculpture

During the 1950s and 1960s, a particular wave of activity across the experimental musical practices of composers including David Tudor, Alvin Lucier and Steve Reich among others influenced by John Cage, reflected a particular interest in foregrounding the objects, materials and technologies of sound-making in ways which were often kinetic or sculptural in some way. Tudor's *Rainforest* (in various versions since 1968) appropriated found objects and materials such as bed springs as resonant 'sculptural loudspeakers' driven by unique hand crafted electronic audio devices. Lucier's *Music for Solo Performer* (1965) used loudspeakers wired to equipment for amplifying brainwave activity to rattle percussion instruments in a collaborative performance between the human and the technological, and in the piece *Pendulum Music* (1968) Reich swung microphones as pendulums above loudspeakers to which they were connected. The resulting bursts of howling feedback are rhythmically modulated by the swinging, gradually become longer and more drawn out as the pendulums slow down and ultimately come to rest signalling the end of the piece.

These pieces are performed and arranged as much by the objects, materials and technologies of sound making as they are by any human performers. Such musical endeavours are well summarised by the artist, technologist and scholar Nicholas Collins when he describes work of this nature as existing 'in the twilight zone between a concert and an installation' (2007, p. 46) and embodying an ethos of 'music implicit in technology', which he identifies as an approach which served as a paradigm for much American electronic music of the 1970s (ibid). This transitional nature between concert and installation (or sculpture) is further illustrated by some of Lucier's slightly later works, *Music on a Long Thin Wire* (1977) and *Music for Pure Waves, Bass Drums and Acoustic Pendulums* (1980): both of which were conceived as performer-less, gallery type pieces as well as performance pieces for the concert hall.

Concurrent with many of these musical examples, the world of sculpture was exploring the possibilities of sound alongside experiments in energetic transformation and movement in a range of practices broadly defined as kinetic art. A key figure in these developments was the Swiss kinetic sculptor Jean Tinguely (1925–1991) for whom sound had been a key concern since at least the 1950s. In his *Meta Mechanical*

Sound Reliefs (1955) Tinguely produced sound through the kinetic striking of saucepans, jars, glass funnels and wine glasses by light hammers (Hulten, 1987, p. 28). His series *Radio Sculptures* (1962) consisted of deconstructed but operative radios emitting live broadcast sound which was kinetically modified by electromechanical devices fitted to their tuning dials and volume controls. In later developments of the *Meta Mechanical* works Tinguely developed the much larger *Meta-Harmonie* series (1979–1985). These industrial-sized assemblages of steel cogs and wheels, wires, belts and musical instruments emit a cacophony of mechanical noise alongside occasional pitched or percussive sounds.

Other kinetic sculptors also introduced sound to their work during this time. Takis worked with magnetism, electromagnetism and movement to create sound pieces such as *Magnetic Pendulum Musical* (1965), *Telemagnetic Musical* (1966) and *Electro-Musical Relief* (1966), all of which appropriated the kinetic, sound-making possibilities of electromagnets, fixed magnets and resonant wires under tension, sometimes strung over a taught canvass working as a sound board. Len Lye, whose 1960s oeuvre is also recognised as sonically inspired kinetic art (see Wall, 2018) similarly worked with magnetism, movement and sound in *The Loop* (1963, also titled *Universe*). This is a 22-foot strip of polished steel formed into a band, which is both tethered and energised by a strong electromagnet inside a plinth. The steel emits tones and harmonics as it rocks and wobbles around, occasionally lurching up high enough to strike a ball suspended above it, exciting a further set of harmonic behaviors within the band.

By crossing the genre divides between time-based music and static material sculpture, these examples help to define the multidisciplinary mode of creative practice which is kinetic sound art. There is more which unites these examples than just kinetic movement and sound however. Themes of energetic relations and material translations are foregrounded within the pieces by presenting shifts that occur between material, electricity, magnetism, physical force, sound and human interaction.

The pieces are also concerned with presenting systems of unpredictable and complex behaviours through the creation of active technological assemblies. These assemblies favour unknown outcomes over fixed objects or structures in an approach neatly summarised by the art critic Jack Burnham as artists aiming to create a 'situation in which things can happen rather than an object per se' (Burnham, 1968, p. 271). Working artistically with unknown behaviours in this way was something that John Cage had previously explored in his indeterminate compositions, as had Umberto Eco in his 1959 essay, 'The Poetics of the Open Work', where he identified that in such creative methods an 'incomplete knowledge of the system is … an essential feature' (Eco, 2004, p. 171). A particular feature of kinetic sound art is that the unknown and unexpected is played out technologically, materially and kinetically, in real time and in the moment of the audiences' reception of the work. David Toop effectively summarises this aesthetic nature of kinetic sound art when he describes how 'as resonant or amplified solids move and interact, activated by unpredictable systems, the patterns of sound they create take on the drama of natural emergent phenomena' (Toop, 2002, p. 125).

Systems and Material Systems

Themes of technological interactions and translations and of complex behaviours are what tie the world of kinetic sound art to the broader field of systems art and systems thinking of the 1960s. Emerging from the scientific world of the early twentieth century the notion of systems thinking became a far reaching cultural and technical influence that made its way into diverse areas such as military defence, urban planning, technological design and computing, art and music. Art critic Jack Burnham, often credited with providing the most thorough account of systems art of the late 1960s, traces the emergence of systems theory to the biologist Ludwig von Bertalanffy (Burnham, 1968, p. 317). In contrast to the polarisation of vitalistic views of the world that emphasised notions of life force and more mechanistic materialist positions based in the physical sciences, Bertalanffy's systems perspective took a line which focused on themes of relationships between components, complexity, multileveled organisation and behaviours (see Skrebowski, 2006). Related areas of development within the sciences accompanied Bertalanffy's general systems theory, including cybernetics, information theory and game theory. From a cultural perspective, in 1968, Burnham enthusiastically pointed out that:

> For the first time in history our culture has the option of literally fusing organic activities with the linear-geometric precision of machines. This has come about recently by the growth of both theoretical and applied sciences: cybernetics, electronic circuit theory, information theory, systems analysis, etc. (Burnham, 1968, p. 69)

With an emphasis on relationships rather than things, and in the context of a burgeoning world of computer and software technology, systems art, particularly through Burnham's critical lens, became increasingly concerned with more abstracted approaches such as conceptual and data-driven artworks and digital technologies. This can be seen as reflective of a broader tendency within systems influenced thinking of the 1960s which was to conceive of and model the natural world through mathematical formalisation, computing and technology.

Increased computer power was a result of the technological systems approach of packing complex circuitry inside ever smaller closed boxes (from mainframes to sub-assemblies to integrated circuits), before treating and arranging those closed boxes of complex relationships as single elements in new relational assemblies. This conceptual and practical method described by the systems-conscious term 'black-boxing', enables the building of ever more complex and powerful systems. Computing technology contributed to and benefitted from such processes. Multimedia artist and writer Casey Alt points to a moment in the 1960s when, with the rise of object-orientated programming languages, the computer system became a medium in its own right aside from traditional conceptions of hardware media objects or materials such as punch cards or magnetic tape. According to Alt, this is the point at which the system itself – the nature of the relationships – becomes the medium, a situation commonly identified today through the general term 'digital media' (Alt, 2011).

Burnham took his conception of systems art along a parallel route to these technological developments and – pre-empting the age of digital media – curated the exhibition *Software* at the Jewish Museum in New York in 1970. The exhibition was 'predicated on the idea of software as a metaphor for art' (Shanken, 2012, p. 51), exploring information technology and public interaction with information processing systems. It featured pieces by one-time kinetic artist Hans Haacke including *Visitors Profile* (1969), which presented a statistical profile of visitors who had input their details via an input/output terminal, and *News* (1969), a real time teletype print-out of news feeds being circulated to clients of various news services.

Technological, data-driven and relational themes such as these were presented across the exhibition by a range of artists including the conceptualist Joseph Kosuth and the performance artist Allan Kaprow. With these connections in mind, it is also possible to map Burnham's systems art onto other art world tendencies of the time such as conceptual art, 'dematerialised' art – as described by the critics Lucy Lippard and John Chandler (1968) – and 'post-formalist' art (Shanken, 2012, p. 52).

Kinetic sound art along with kinetic art in general was both critically subsumed into and overshadowed by this broader debate around systems and systems art during the latter part of the 1960s and into the 1970s. In comparison to art work, which was responding to themes of data networks and computing technology, the world of kinetic sound sculpture with its emphasis on objects and movement and with a lineage connecting it to mechanical automata and Dadaist sculpture began to look decidedly outmoded in the eyes of critics. Scholar Christina Chau argues that as a prominent critic of the time, Burnham ultimately tried to 'sequester the theory and practice of movement in art away from postmodern aesthetics' (Chau, 2014, p. 64) describing kinetic art as 'unrequited'. Vadim Keylin similarly claims that much sound sculpture is excluded from sound art discourse for being 'so unmodernly modernist' and thus remains underexplored (2015, p. 182).

As such, identifying the particular aesthetic concerns of kinetic sound art involves recognising the systems-conscious relational features of the work alongside and as equal to its material- and object-based nature. The systems of kinetic sound art are localised and materially embedded rather than conceptual, displaced or abstracted. In the kinetic pieces described, the technological translations at play are evident at the time and in the place that the work is received: motors spin, pieces of wire resonate, objects move and vibrate, pendulums swing, electricity flows and sound is audible all in the moment. It is these concerns within kinetic sound art that are identifiable as 'material systems', in contrast to the more abstract, data-influenced and conceptual approaches typically emphasised by the discourse around systems aesthetics and systems art.

Material Systems and STS

Approaches and sensibilities from STS contribute in a number of overlapping ways to the understanding and practice of kinetic sound art. Firstly, STS is useful in developing the idea of the material system which has been shown as critical in defining kinetic practices against the larger swathe of systems sensibilities within the arts. The key to this is neatly summarised by STS scholar John Law's description of the STS actor network approach as a 'sensibility to the messy practices of relationality and materiality of the world' (Law, 2009, p. 142). It is Law's conjoining of both relations and materials which clearly creates a parallel concern to the material systems of kinetic sound art.

The STS field is rich with detailed examples of these relational, material systems, often crossing over between human and non-human entities, emphasising sociotechnical relations. Bruno Latour's descriptions of the development of technologies such as the Kodak compact camera (1987, pp. 115, 131) and Thomas Edison's electric light (ibid., p. 239) are two such examples.

Latour presents the Kodak compact camera as a system of wood, steel, photosensitive chemicals, celluloid and so on, which needs to act as a single entity in a wider, commercial market, itself a system of sales outlets, film distribution and film-developing services. This example is a good illustration of how the systems-conscious notion of black-boxing serves STS accounts well. In Kodak's commercial technology the messy material business of film developing is taken care of by their film-developing service, it is black-boxed both physically by the camera and by Kodak's network of technical services. Similarly, in his account of Edison's development of the incandescent light, Latour gives equal attention to materials and their relations. The electric light is presented as a system whose commercial success is predicated on such material concerns as the resistance of the light filament and the cross-sectional diameter of cables as well as more sociotechnical concerns, such as the price of gas and copper.

Drawing on work by Latour and Woolgar, Law develops these themes of material practices and relations becoming black-boxed, through an example from laboratory science, the mass spectrometer. Law shows how this pre-packaged standardised piece of laboratory equipment 'incorporates the majority of an earlier body of scientific activity' (Law, 2004, p. 33) and as such bundles together machines, skills and statements into a reified black box that outputs graphical representations of data extracted from material samples. Law describes how using this pre-packaged system of mass spectrometry is akin to buying a personal computer rather than understanding the electronics embedded within and assembling one out of components (ibid., p. 33). The mass spectrometer, like the personal computer, is a machine but also a set of statements, assumptions and applications which, as much as possible, have been fixed into an established method of scientific activity. Law shows how such fixed and standardised machines and methods are typically used as 'inscription devices' (ibid., p. 20), translating the behaviours of various materials within the laboratory into neat, comprehensible graphs or sets of data, which can easily be presented or published as scientific texts and papers. In this way, Law points out, the materiality of the scientific process gets deleted in what is ultimately circulated in a textual format (ibid., p. 20).

The same awareness of this shift between material activity and comprehensible data is discernible in Latour's discussion around 'centres of calculation' (Latour, 1987, p. 215). Here Latour explores how Ohm's law is an abstraction of the kind of problems Edison was trying to resolve in commercialising his incandescent light bulb. Latour identifies this single abstraction of a particular set of material relationships as the beginning of an ascent into higher and higher order abstractions which eventually become associated not with materials and not even with material processes or their production, but with theories in the mind of the producer (ibid., p. 241). For Latour, a single localised abstraction of a particular material interaction represents a recognisable relation to a material world, but it also opens the door to an ascent of more removed, more generalised abstract concepts.

The shifts from material relations to textural meaning and from localised material activity to generalised abstract concept that are highlighted within these STS accounts are useful in identifying parallel concerns for kinetic sound art. STS accounts draw attention to material systems hidden inside equipment and embedded within scientific concepts and holds them in mind whilst following the wider activity and impact of a technology or a scientific practice. Similarly, kinetic sound art explicitly presents material relations and activities as part of the work such that an audience is conscious of them during the work's reception and as part of its impact. Kinetic art practices pull back from allowing those relations to become abstracted beyond the immediate, localised materials and sounds of the work. The impact of kinetic sound art is not typically or easily translated into a textual understanding or an abstracted concept, it remains embedded in the active material relations at play in the work. In this way kinetic sound art invites a sensibility equivalent to that presented through STS accounts. This critical feature of much kinetic sound art may have ultimately led to its underrepresentation by an art world enthused by the possibilities of abstracted, textual and globalised data systems and, as such, STS accounts are useful in unpicking and identifying these important themes for the genre.

Artists Constructing Technology

Aside from helping to identify and understand the critical theme of material systems within kinetic sound art, STS provides a useful framework for understanding more generally the activities of artists and composers who use and assemble technologies as part of their creative practice as so many kinetic sound artists have. These insights also stem from the idea of opening the technological black box of commercial technology and are also neatly illustrated through Latour's study of the Kodak compact camera. Latour identifies how semi-professional photographers chose to open their cameras and do their own film-coating and developing, unlike amateurs who simply used compact cameras and made use of Kodak's commercial film-developing services. This represents a deeper level of engagement with the material process and as such the creative process of photography for the semi-professionals. They choose to tamper with the photographic technological black box as a part of their creative practice. The same motivation is evident with sound artists wanting to explore the possibilities of opening up and reassembling elements of commercial sound making technologies such as loudspeakers, radios and microphones. The pieces described by Lucier, Tinguely, Tudor, Takis and others do exactly this by either arranging existing technologies in novel configurations, by remaking technologies from a component level or by modifying and reworking existing technologies.

Such artistic methods which rework existing technological arrangements to explore new possibilities and critique the world of commercial sound technology have been explored from an STS perspective through contemporary artistic practices such as 'hacking' and circuit bending (see Pinch, 2016; Parikka & Hertz, 2012). These practices emphasise a non-expert approach to tinkering and intervening in the black boxes of sound technology to explore new sounds by making untested and unknown connections and contacts on circuit boards. This creative approach belongs to a lineage of practice which also contains a wider breadth of technologically astute artistic activities and where kinetic sound art is an important feature. The sound artists discussed here for their kinetic works all contribute to this lineage of practice, with David Tudor being a particularly notable figure in the assimilation of technical knowledge into artistic methods.

Tudor's initial foray into working with sculptural speakers of the type used in *Rainforest* was as part of the ambitious *9 Evenings of Theatre and Engineering* event in New York in 1966. This event included interactive performances and electronically modified musical instruments, amplified body movements and brain-wave activity, and robotic machines among other technological endeavours from artists including Alex Hay, Robert Rauschenberg, John Cage and Tudor. Overseeing the engineering complexity of the event was Bell Laboratories physicist and systems engineer Billy Klüver. Klüver had already acted as advisor for many kinetic artworks and exhibitions

(Burnham, 1968, p. 281) and had in 1960 been introduced to Cage and Rauschenberg by Tinguely. In the year previous to the *9 Evenings* event he had helped in the realisation of Cage and Merce Cunningham's electronic composition and performance piece *Variations V*, along with another Bell Laboratories engineer Max Matthews (Miller, 2001, p. 552). As a systems engineer with a busy work schedule within the kinetic and new media arts of the time, Klüver commented on the inevitability of a closer relationship between artists and engineers emerging (Burnham, 1968, p. 281). Burnham similarly identified a reliance on technical specialists by artists in the mid 1960s, commenting on a 'new and growing symbiosis' (ibid., p. 281). For Tudor, following the *9 Evenings* event and his experiences of working with engineers like Klüver, this symbiosis moved beyond a reliance on the services of engineers and into his own assimilation of engineering knowledge and skills. He went on to further develop his sculptural speakers and also embarked on designing and fabricating his own hand-built bespoke electronic audio devices that would become a characteristic feature of his work, establishing him as a 'soldering composer'.

Although taking materials and techniques from commercial technologies and hobby electronics resources of the time, Tudor's approach was more than simply the hacking and the non-expert undoing of existing technologies. Along with other artists such as Gordon Mumma who used technology to create kinetic and electronic sound art, Tudor's work evidences a keenness to assimilate engineering knowledge directly into artistic outcomes. Such an approach requires an individual to act as both an engineer and an artist, balancing and merging the different knowledge, skills and priorities of each field. Other examples of this more involved and evolved relationship with electrical and mechanical technology within contemporary sound-art practice include the work of Godfried Willem Raes' Logos Foundation, Trimpin and the artists Peter Bosch and Simone Simons, all of whose kinetic sound sculptures and machines, designed and built by the artists, are often notable engineering achievements.

From these component level artistic constructions through to the novel arrangement and foregrounding of existing black-boxed technologies (such as microphones and loudspeakers), it is possible to conceive of a general field of the artistic construction of technology for which kinetic art and sound art are key elements. This field is obviously large and has permeable boundaries but perhaps one defining feature is that the technology being constructed is not intended to have commercial or scientific viability in the standard way. Its successful dissemination is achieved through artistic activities of performances and exhibition. While these activities could themselves be explored from a sociotechnical perspective, the effective success of artistically constructed technology is not predicated on its ability to maintain a fixed techno-scientific reality but on an artistic reality: a sonic or visual structure, a mood, an idea. As has been shown, often this artistic reality can involve evoking the very opposite state to techno-sciences standard aim of controlled repeatability. Within the material systems of kinetic sound art for example, unknown and unknowable chance events and behaviours are often favoured by artists whose aim is to create a 'situation where things can happen rather than an object per se'.

Questions arise however around what 'things happening' may be considered successful or unsuccessful for a particular artwork. Unknown events may be considered acceptable within certain parameters – new and unexpected sound events for example – but perhaps not to the point of an artwork appearing broken or non-functional in some way. In this regard kinetic artwork does need to stabilise to some extent for exhibition purposes, at least to the point where the artist/engineer, audience and curator are satisfied by the type of unpredictable behaviour exhibited. This represents a tension between the freedom of artistic expression and exploration and the need for technology to take on a fixed nature of reliable operation, which all technologically engaged artists will have grappled with. This tension is well illustrated by Trevor Pinch (2016) in his discussion of a slightly broken synthesiser module that makes interesting sounds but cannot always be wholly relied upon in performance. Such questions around predictability and brokenness may also be arguably present in some of Tinguely's self-destructing artworks such as *Homage to New York* (1960) and *Study for an End of the World* (1962).

Electromagnetic Interrogations

Electromagnetic Interrogations (2011–2014) is an original piece of kinetic sound art made by the author, that has been exhibited a number of times in the UK, including at the Kinetica Art Fair in London in 2017. The piece is a kinetic exploration of material systems that delves into the dematerialised and unseen nature of electromagnetism. It does this by interrogating the electromagnetic radiation given off by three dismantled but still operative compact disc (CD) players, mounted upon three plinths (Plates 4 and 5). Coil microphones, which work as electromagnetic pick-ups, are suspended in a pendulum configuration such that they can swing above the various internal parts of the players including the circuit boards, the motors, the transformers and the display units (Plate 6). The coil microphones (sold commercially as telephone bugs or pickups) transduce the electromagnetic energy that surrounds the various electronic components into an electrical signal, which is then amplified and made audible through loudspeakers.

The actual CDs being played by the players are blank, and hence the only sound made by the piece is a result of the interaction between the electromagnetic activity of the players and the movement of the coil microphones, which are swung by servo motors at timed intervals, periodically coming to rest for a short time. The movement of the pendulum microphones modulates the electromagnetic interference in different ways, ranging from short rhythmic bursts to longer and slower undulating variations depending on how energetically they are swinging. The cycles of events of both the swinging pendulums and the CD players' operations (play, pause, track skip and so on) are unsynchronised, creating a complex soundscape of forever shifting relationships and unexpected events. The coil microphones are amplified individually through their own loudspeakers which vary in shape and size adding further sonic character and spatial depth to the sound produced.

Film documentation of *Electromagnetic Interrogations* (see Pigott, n.d.) gives an overall sense of the different sonic arrangements that can emerge from the piece. A central feature is the low frequency drone given off by the mains power transformers of the CD players whose main harmonic components are defined by the frequency of the national grid at 50Hz (in the UK), though harmonics of this frequency, especially 100Hz, are also present. These low-frequency tones are either rhythmically modulated into short bursts by the energetically swinging microphones, or they become more continuously audible when the microphones are slowing down and coming to rest. The result of these different kinetic possibilities are shifting rhythmic patterns that emerge from the relationships between the different pendulum microphones. The motors of the CD players emit a range of higher frequency sounds including whining and whistling sounds, sometimes ascending in frequency. The optical lasers of the players present a good deal of electromagnetic noise as they track the data on the CD media. This ranges from a broadband white noise present when the CD is playing and lots of data is being read through to 'chirrups' that occur as the laser is searching for a particular track or a rhythmic ticking caused during the fast-forward or reverse function. The fluorescent display units of the CD players and their associated signal paths and ribbon cables emit a characteristic high frequency whine.

Electromagnetic Interrogations is an unpacking of a commercially produced technological black box. The piece draws attention to the internal material relations of the CD players which are exposed and interrogated electromagnetically and sonically. Whilst it is interesting to make the associations between particular sound events and electrical and mechanical events within the CD players it is of course not possible to know exactly what causes every sound that may occur. Many of the sources of electromagnetic radiation are hidden away in the electronics of the circuit boards and integrated circuits – black boxes inside black boxes – and the exact nature and cause of the activity at this level is not really knowable in practical terms. It is interesting to note that even similar functions can sound quite different between the different makes, models and electronic architectures of the particular CD players. The operation of these domestic hi-fi units has been black-boxed by each manufacturer in a slightly different way, creating a slightly different electromagnetic field and sound.

The world of digital audio can be seen as a particular arrangement of social and technological relations, a method for the commercial distribution of fixed media sound files. *Electromagnetic Interrogations* unpacks this arrangement with regards to the compact disc players and exposes some of the material systems necessary for this technology to operate. The players have not been 'hacked' as such – they are fully operative and working as they were always intended to by the manufacturer, but they have been carefully removed from their outer casing – stripped of their outer black box. A second material system consisting of coil microphones is also in operation, interrogating and further exposing the inner workings of the CD players by way of their electromagnetic radiation. From an STS perspective, this second system is a kind of inscription device – measuring and interrogating the material behaviours of the CD players. But this inscription device is not one that is translating between material events and graphical plots, text or data. Rather it is inscribing non-discursive sound and noise whose complex patterns only begin to hold meaning in direct relation to the kinetic material system which produced them. In this way the system is inherently connected to its localised material behaviours.

Furthermore, the piece aims to foreground the fact that the sound produced is a result of an interaction between the CD players and the swinging microphones which only hear and amplify particular parts of the system at particular times. The microphones are not fixed and transparent in their interrogation of the electromagnetic fields, rather they are performative and active in their listening. This embodies the STS theme of inscription devices such as the mass spectrometer and other technologies used in techno-scientific methods, appearing as fixed clean and objective black boxes but actually working as more contingent assemblages, more tentative, unfolding and ad hoc than they may initially appear to be (Law, 2004, p. 41). As with many examples of kinetic sound art the tentative, unfolding, ad hoc and performative nature of the material systems in *Electromagnetic Interrogations* are foregrounded.

Of course, the system must operate as intended: the coil microphones must swing and amplify the electromagnetic waves they encounter, and the exposed CD players must spin and play the silent CDs. On occasion the players themselves have ceased to operate often due to the fact that they have been extracted from their protective black boxes and been laid out at reliefs on a plinth. In such situations the players have been replaced as they are cheap, easily available outmoded technology not well designed for user (or artist) repair.

Conclusions and Assemblage

Law further expands the theme of techno science's black boxes drawing together combinations of ideas and technologies into single methods and tools. He presents a case for thinking of research method as a more open and creative tool for crafting, resonating with and amplifying particular versions of the world (Law, 2004, p. 144). This more open view of method as a process-orientated, rather than an objective- or end-orientated approach to research, Law terms 'method assemblage' (ibid., p. 13). The idea of an 'assemblage' he defines as

> like an episteme with technologies added but that connotes the ad hoc contingency of a collage in its capacity to embrace a wide variety of incompatible components. It also has the virtue of connoting active and evolving practices rather than a passive and static structure. (Law, 2004, p. 41)

This characterisation of a method assemblage describes well the process of investigation into kinetic sound art and STS discussed here. By combining a wide variety of components including art historical accounts and critiques, STS, motors, amplifiers, systems theory, Ohm's law, listening, magnetism, reading, reflection and writing, the investigation is not a linear one as a written account such as this might imply. The investigation is an assemblage that has developed through the activities of creative practice and research, moving back and forth between potentially incompatible components. Resonant themes between STS and kinetic sound art have been found, crafted and amplified both through the practice of kinetic sound sculpture and through formulating written accounts. The full range of these activities form the method assemblage of the project. The materiality of this process is preserved in the piece *Electromagnetic Interrogations* and disseminated through exhibition and to an extent through video documentation of the work, whilst written accounts translate the same themes through other networks.

Through these different possible channels of translation and dissemination the themes remain the same, resonating across STS and kinetic sound art. They include a sensibility to the material systems of technology such that they are not subsumed entirely into data systems of purely textual abstracted meaning. And a context of artists constructing technology, sometimes to a high technical standard of engineering but towards outcomes which are not simply trying to fix a techno-scientific state. Rather, these outcomes often aim to evoke an alternative view of technology as a contingent and evolving system. For these artists there is also a tension between communicating this contingent nature of technology and producing technological artworks that will reliably 'perform' and work in the gallery or concert hall. Finally, the same themes allow for a reflection on the assemblage nature of methods for simultaneously making connections, insights and artwork. Whether the outcomes be written accounts or material sculptures, they are made as ad hoc combinations of apparently incompatible components, combining ideas and materials equally in an evolving material system.

References

Alt, C. (2011). Objects of Our Affection: How Object Orientation Made Computers a Medium. In E. Huhtamo & J. Parikka (Eds.), *Media Archaeology: Approaches Applications and Implications* (pp. 278–301). Berkeley and Los Angeles: University of California Press.

Burnham, J. W. (1968). *Beyond Modern Sculpture*. London: Penguin Press.

Chau, C. (2014). Kinetic Systems: Jack Burnham and Hans Haacke. *Contemporaneity, 3*(1), 62–76.

Collins, N. (2007). Live Electronic Music. In N. Collins and J. d'Escrivan (Eds.), *The Cambridge Companion to Electronic Music* (pp. 38–54). Cambridge: Cambridge University Press.

Cox, C. (2013). Sonic Philosophy. Available at http://artpulsemagazine.com/sonic-philosophy (accessed 8 July 2019).

Eco, U. (2004). The Poetics of the Open Work [1959]. In C. Cox and D. Warner (Eds.), *Audio Culture: Readings in Modern Music* (pp. 241–249). London and New York: Continuum.

Hulten, P. (1987). *A Magic Stronger than Death*. London: Thames & Hudson.

Keylin, V. (2015). Corporeality of Music and Sound Sculpture. *Organised Sound*, *20*(2), 182–190.

Krauss, R. E. (1979). Sculpture in the Expanded Field. *October, 8* (Spring), 30–44.

Latour, B. (1987). *Science in Action*. Cambridge, MA: Harvard University Press.

Law, J. (2004). *After Method: Mess in Social Science Research*. New York: Routledge.

Law, J. (2009). Actor Network Theory and Material Semiotics. In B. S. Turner (Ed.), *The New Blackwell Companion to Social Theory* (pp. 141–158). Oxford: Blackwell.

Licht, A. (2009). Sound Art: Origins, Development and Ambiguities. *Organised Sound 14*(1), 3–10.

Lippard, L., & Chandler, J. (1968). The Dematerialization of Art. *Art International, 12*(2), 31–36.

Miller, L. E. (2001). Cage, Cunningham and Collaborators: The Odyssey of Variations V. *Musical Quarterly, 85*(3), 545–567.

Parikka, J., & Hertz, G. (2012). Zombie Media: Circuit Bending Media Archaeology into an Art Method. *Leonardo Journal, 45*(5), 424–430.

Pigott, J. (2017). Across Fields: Sound Art and Technology from an Electromechanical Perspective. *Organised Sound, 22*(2), 276–285.

Pigott (n.d.). *Sonic Marbles*. Available at www.sonicmarbles.co.uk (accessed 8 July 2019).

Pinch, T. (2016). "Bring on Sector Two!" The Sounds of Bent and Broken Circuits. *Sound Studies, 2*(1), 36–51.

Shanken, E. A. (2012). In Forming Software: Software, Structuralism, Dematerialisation. In H. Higgins & D. Kahn (Eds.), *Mainframe Experimentalism* (pp. 51–62). London: University of California Press.

Skrebowski, L. (2006). All Systems Go: Recovering Jack Burnham's 'Systems Aesthetics'. *Tate Papers, 5*. Available at http://www.tate.org.uk/research/publications/tate-papers/issue-05 (accessed 8 July 2019).

Toop, D. (2002). Humans, Are They Really Necessary? In R. Young (Ed.), *Undercurrents: The Hidden Wiring of Modern Music* (pp. 117–129). London and New York: Continuum.

Wall, S. (2018). Len Lye's Kinetic Experiments: Sounds of Sculpture. *Journal of Sonic Studies*, 16. Available at https://www.researchcatalogue.net/view/457239/457240/0/0 (accessed 8 July 2019).

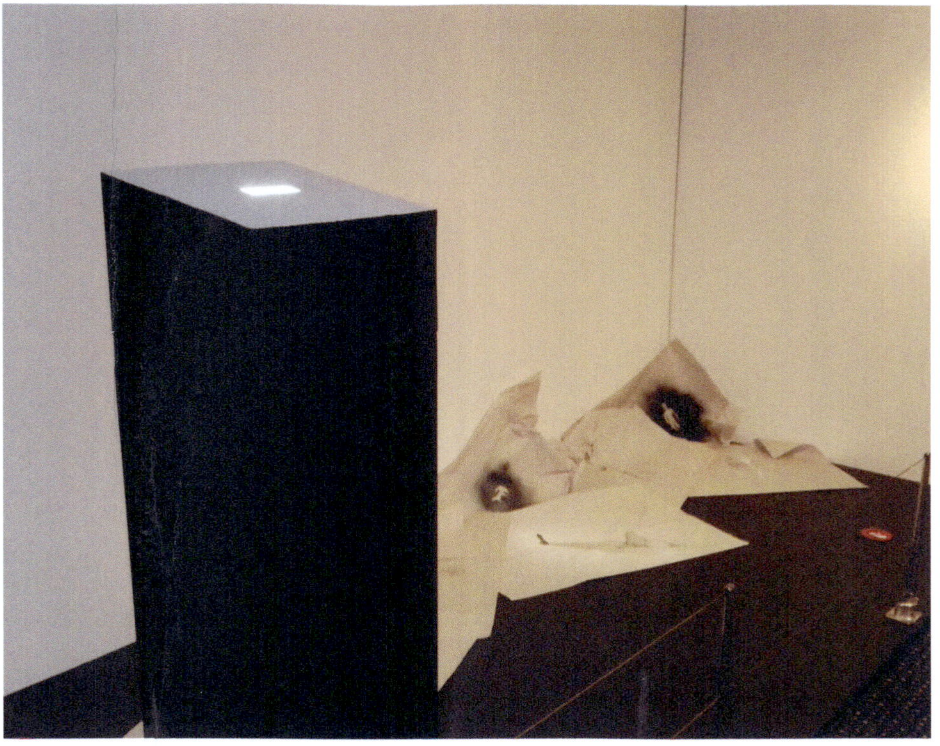

Plate 1 Benjamin Forster, *Kynic* (2013). Photo by Nora Vaage.

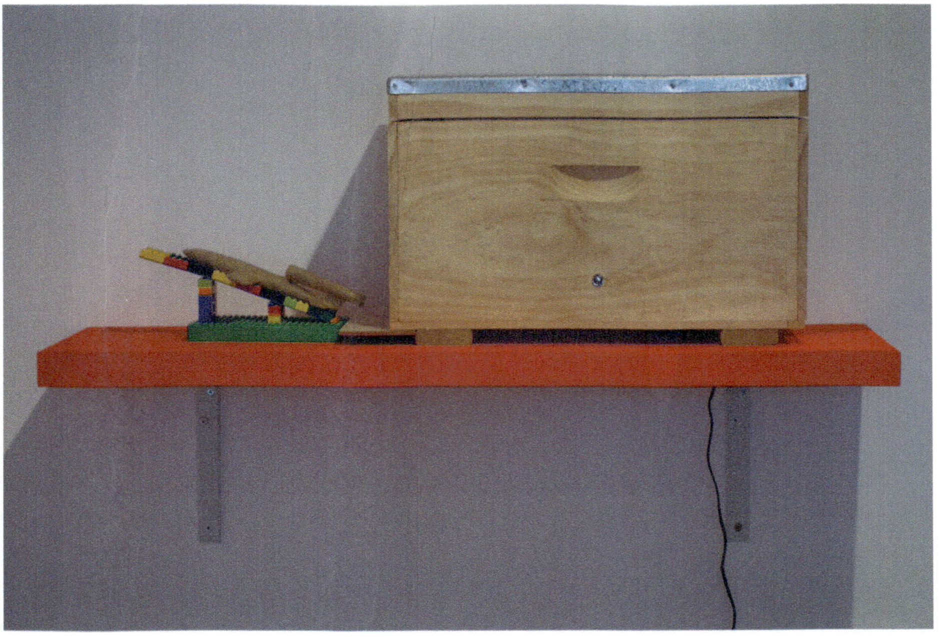

Plate 2 Nigel Helyer, *Float Like a Butterfly; Sting Like a Bee* (2013). Hive with audio actuator and spy-hole lens. Photo courtesy of Nigel Helyer.

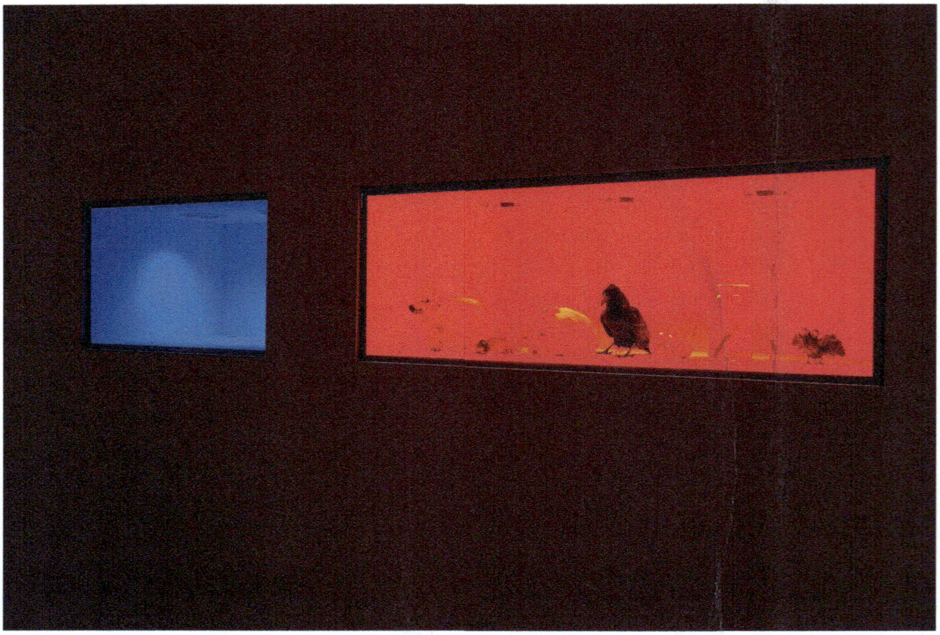

Plate 3 Oron Catts & Ionat Zurr, NoArk II (2009). Tissue Culture and Art Project. Installation photo by Mark Ashkanasy, *Superhuman: Revolution of the Species*, RMIT Gallery.

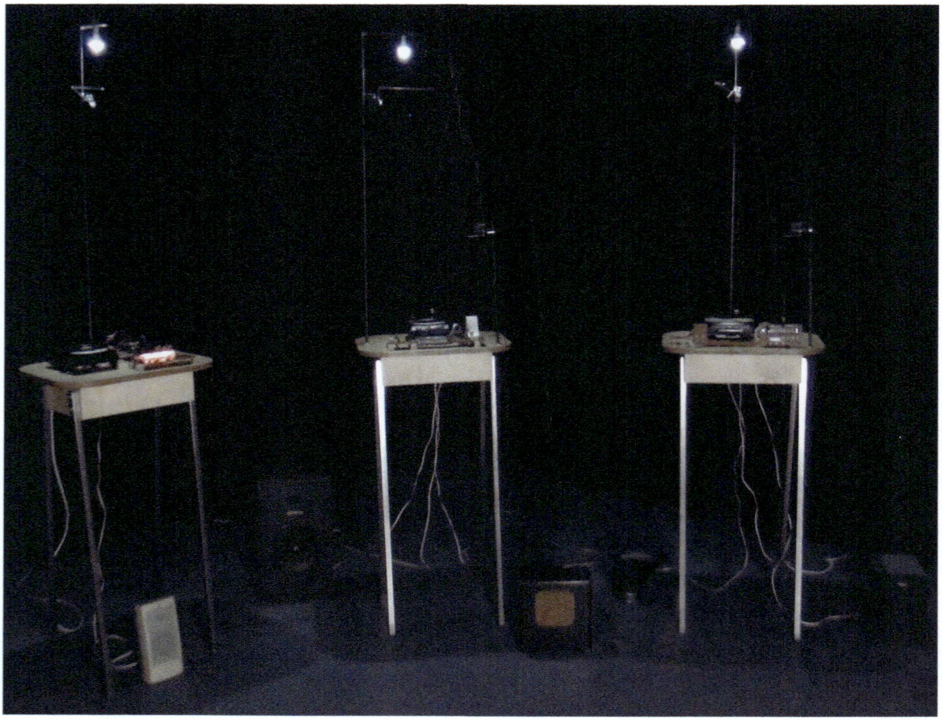

Plate 4 Jon Pigott, *Electromagnetic Interrogations* (2011–2014).

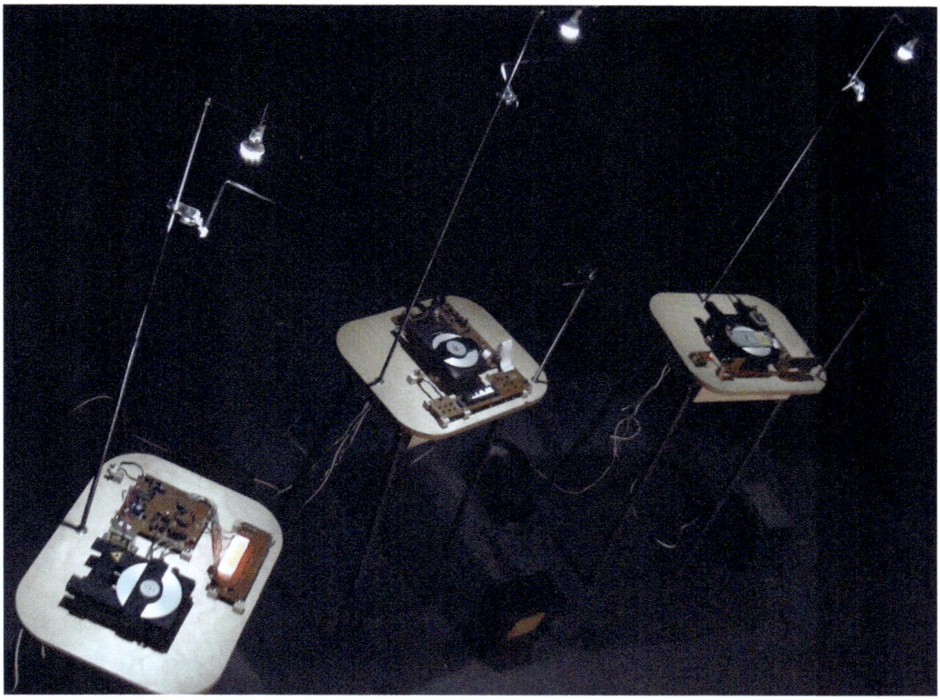

Plate 5 Jon Pigott, *Electromagnetic Interrogations* (2011–2014).

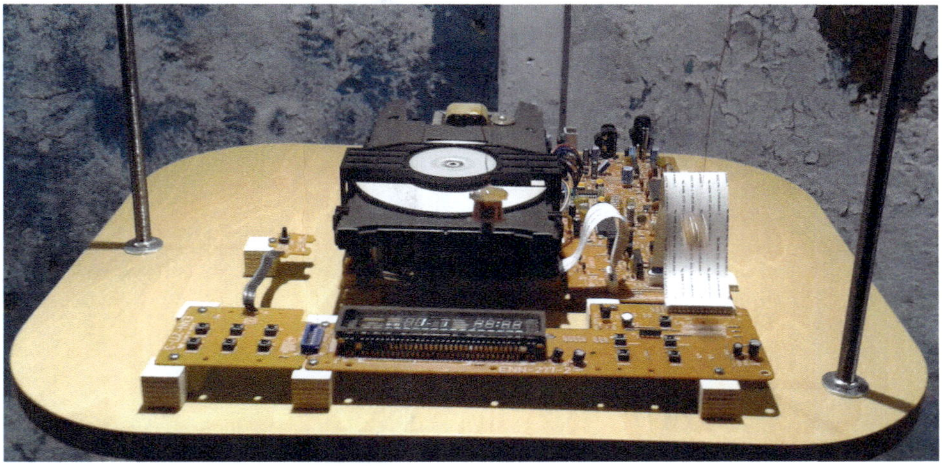

Plate 6 Jon Pigott, *Electromagnetic Interrogations* (2011–2014).

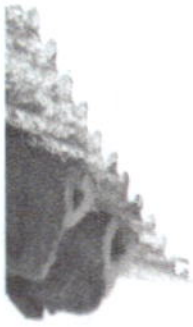

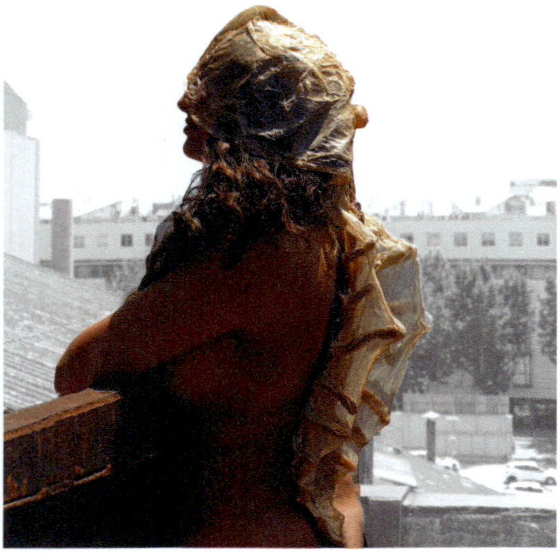

Plate 7 Ingried Ramirez, Jessica Dias, Noor El-Gewely, & Thora H. Arnardottir, *BIO[lum]SKIN* (2016). Students' work at IaaC for studio: SKIN2.

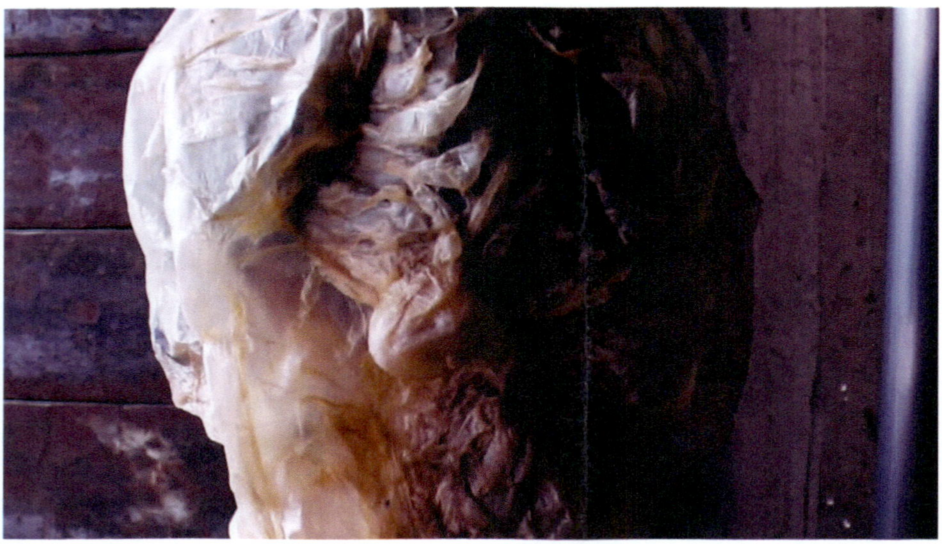

Plate 8 Ingried Ramirez, Jessica Dias, Noor El-Gewely, & Thora H. Arnardottir, *BIO[lum]SKIN* (2016). Students' work at IaaC for studio: SKIN2.

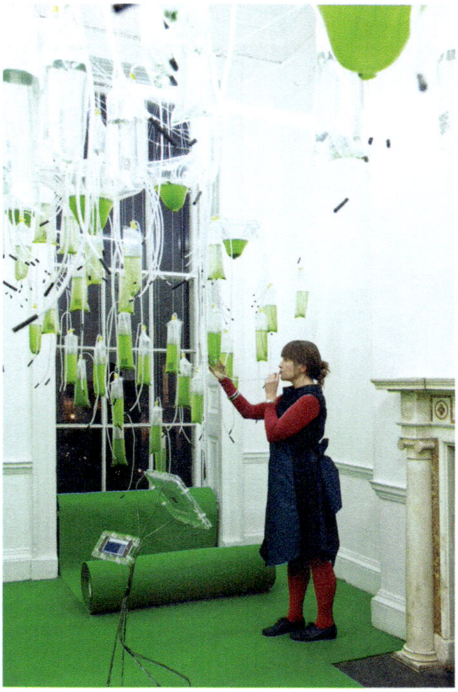

Plate 9 ecoLogicStudio, *H.O.R.T.U.S.* (2012).

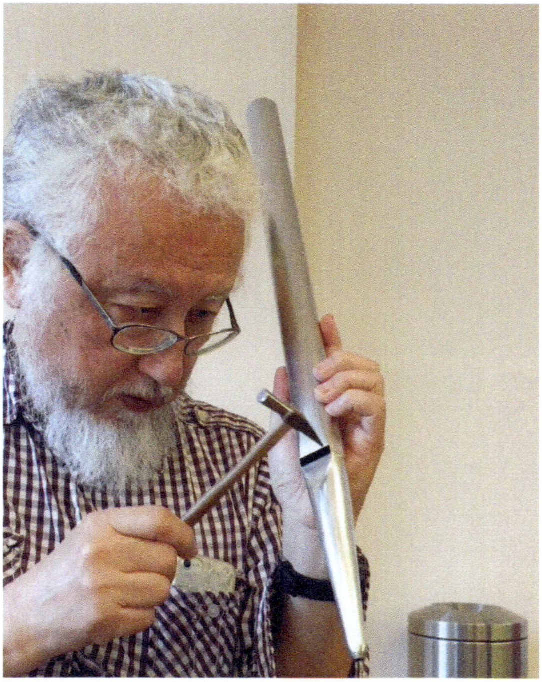

Plate 10 Munetaka Yokota using a hammer during the final voicing of a mixture pipe. Photo by Peter Peters.

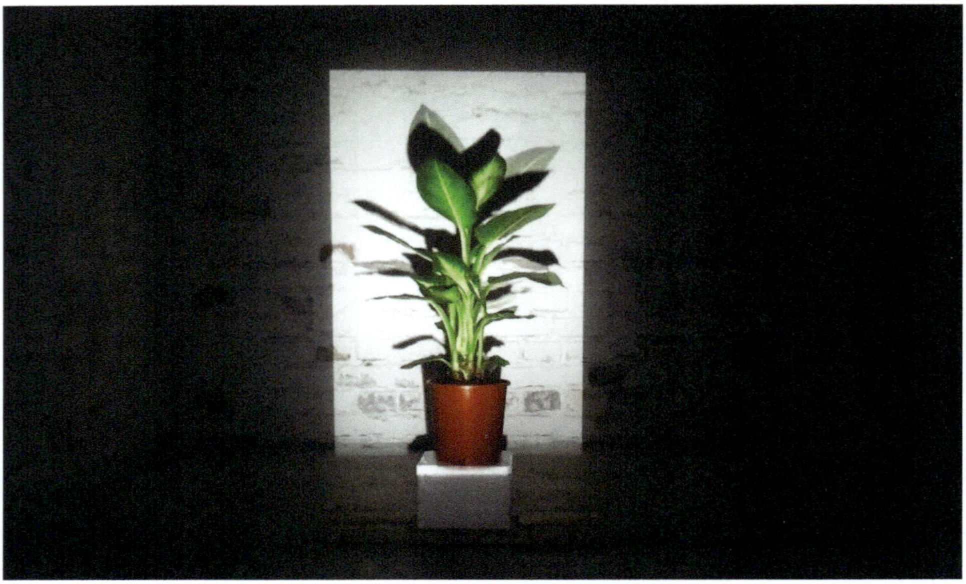

Plate 11 Claude Draude & Marc Horstmann, Everything is Taken (June–July 2011). Mixed-media installation, Kreuzbergpavillon, Berlin.

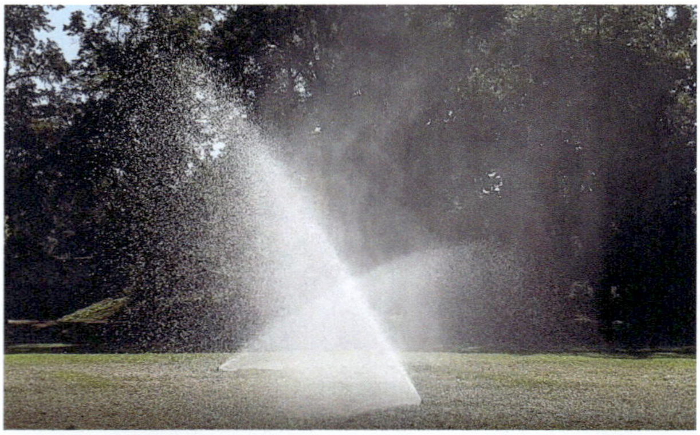

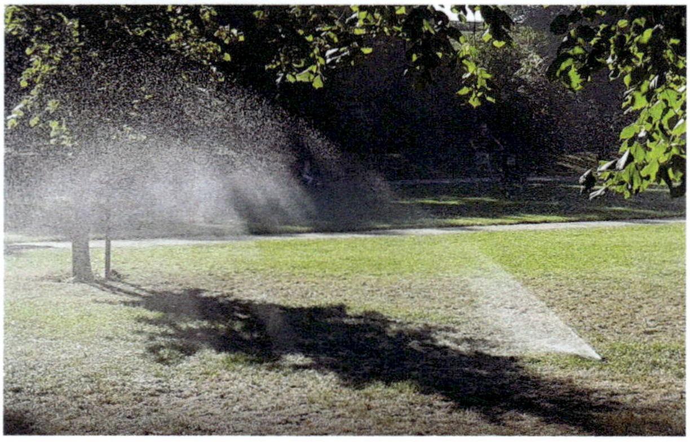

Plates 12a, 12b Claude Draude, Everywhere, Nowhere (Aug. 2016). Field study, Vyšehradské sady, Prague.

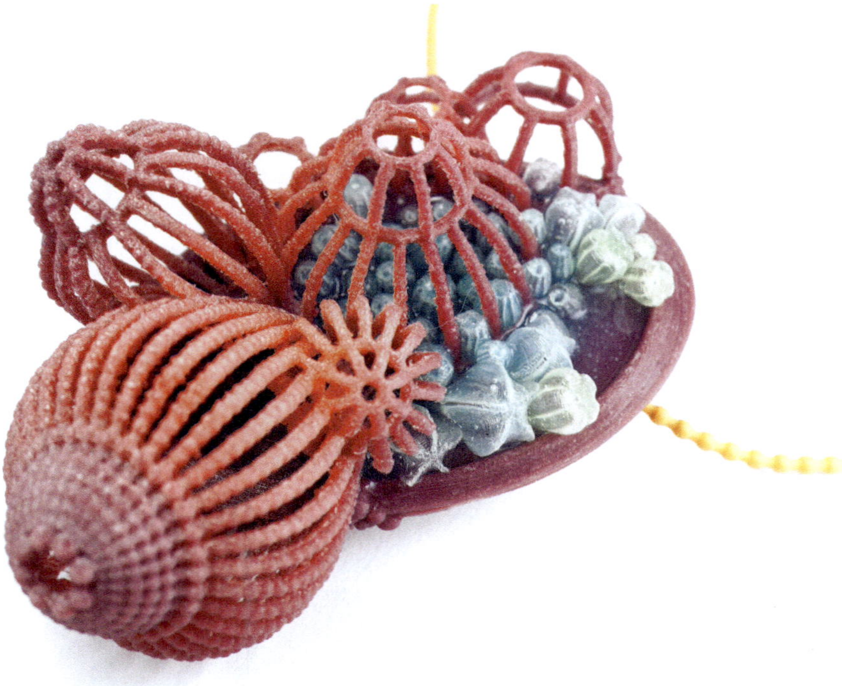

Plate 13 Katharina Vones, Radiolaria Pendant (2017). Thermochromic PLA, photochromic silicone, sterling silver, polyurethane. Photograph by Katharina Vones, ©2017.

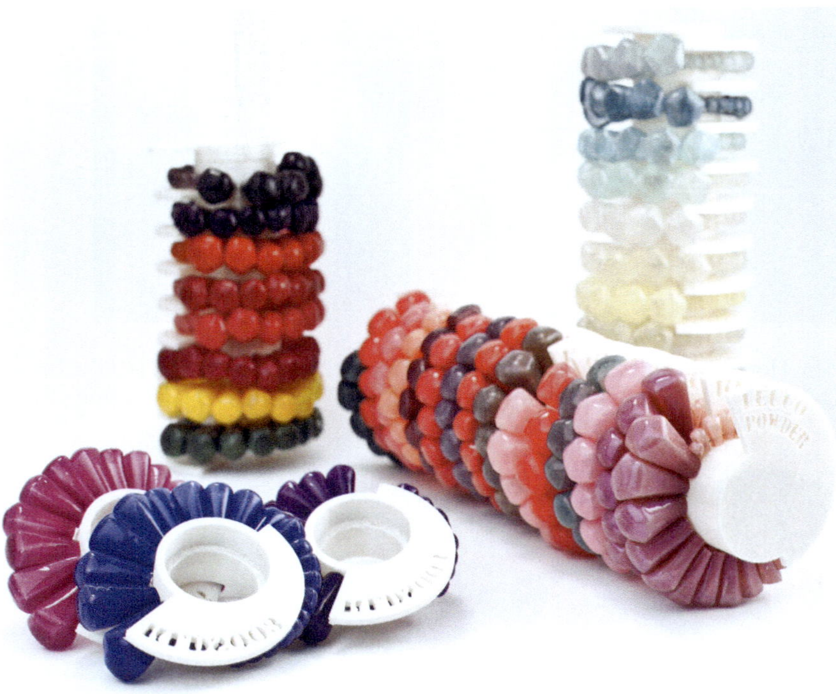

Plate 14 Katharina Vones, personal materials library. Thermochromic PLA and photochromic silicone samples, grouped by type of pigment used and experimental protocol followed. Photograph by Katharina Vones, ©2019.

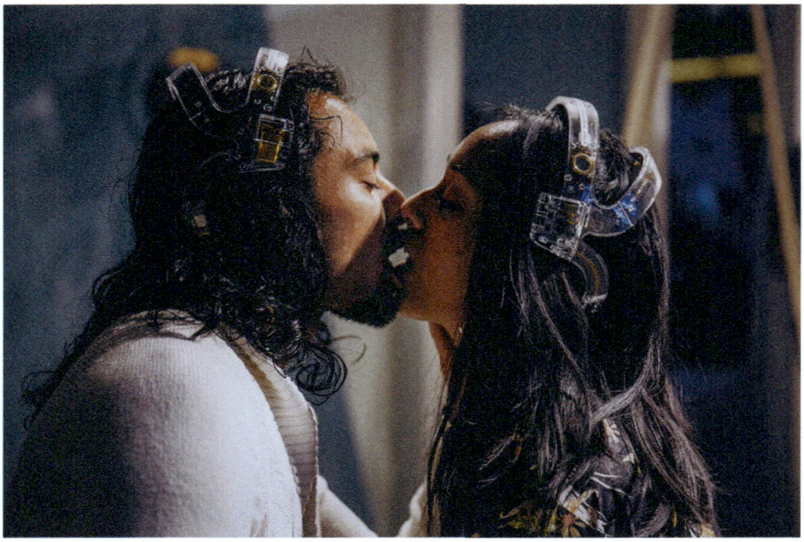

Plate 15 Hermen Maat & Karen Lancel, *EEG KISS* (2016). Image: Max Jansen. © Lancel/Maat.

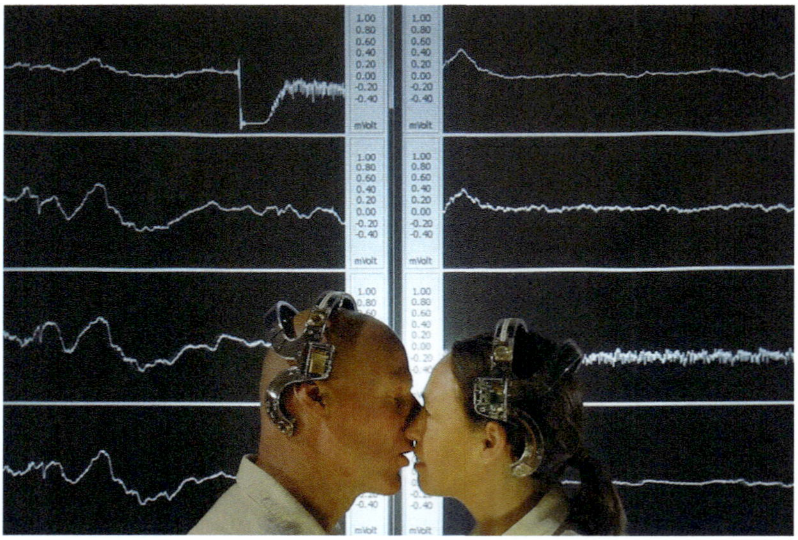

Plate 16 Hermen Maat & Karen Lancel, *EEG KISS* (2014). Film still, Eye Film Institute & Discovery Festival, https://vimeo.com/158299181. Film: Daan Hartoog. © Lancel/Maat.

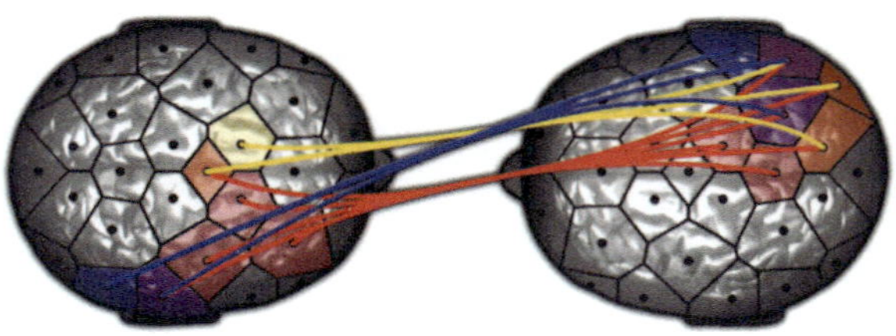

Plate 17 Image of "inter-brain synchronization in alpha (blue), beta (orange) and gamma (red) frequency bands related to interactional synchrony during spontaneous imitation of hand movements" (2011). CC-BY, Guillaume Dumas, Towards a Two-Body Neuroscience. *Communicative and Integrative Biology*, *4*(3), 349–352.

8 Negotiation, Translation, Synchronization?

The Role of Boundary Objects in Artistic Research[1]

Johanna Schindler

In the Field

We are at a Swiss Academy of Art and Design. The artistic research project I investigated here was funded by the Swiss National Fund. The three main researchers—Anna, a literature scholar; Nick, a media artist; and Steve, a computer scientist—were carrying out empirical research on the notion of atmospheres. To this end, they designed and developed a computer-controlled, responsive environment (called "the setting" hereafter) and examined the behavior of test persons inside the environment as well as their perception of it. Each of the three researchers contributed to the design and development of the setting with their specific ideas, competences, and interests. My role as an ethnographer was to observe their daily working practices, routines, and spaces and to carry out interviews with the researchers as well as to assist them with small tasks. I did so three times over the course of one year, each time for three weeks.

During my first day of research, we spent approximately two hours in the "lab," which was a space of approximately 400m^2 and hosted several research projects affiliated with the academy in individual "research islands," working on the setting. This work consisted in cutting several lengths of cloth and arranging them both on a previously installed aluminum frame and on the floor, discussing where and how to put them, and adding further layers to create passages as well as closed spaces. Each of the three researchers contributed their ideas and thoughts on the setting's design. On the second day, Nick and David, a computer scientist and student assistant in the project, developed an initial technical setup, and on the afternoon of the third day, we spent a few more hours on refining the arrangement of the fabric, for example adding several layers of blankets to upholster the floor. In the late afternoon of that day, the setting already resembled the state displayed in Figure 8.1.

Test persons were supposed to enter the setting either from the left or the right side. Before entering the space, they were equipped with a chest belt including two sensors to measure the speed and depth of their breathing as well as their movement. A pulse sensor attached to the finger measured their heartbeat. The data were transferred to a computer in real time through a sender integrated in the chest belt. With the help of mappings previously programmed with MaxPatch, the data provoked changes in the color and brightness of the setting's lighting to mirror the heartbeat, in the motion of a fan to reflect the breathing rhythm and movement visually, and in the activation of a subwoofer as well as wind sounds that imitated the pulse and breathing sonically. The test persons were asked to explore the setting freely, most of them moved around the

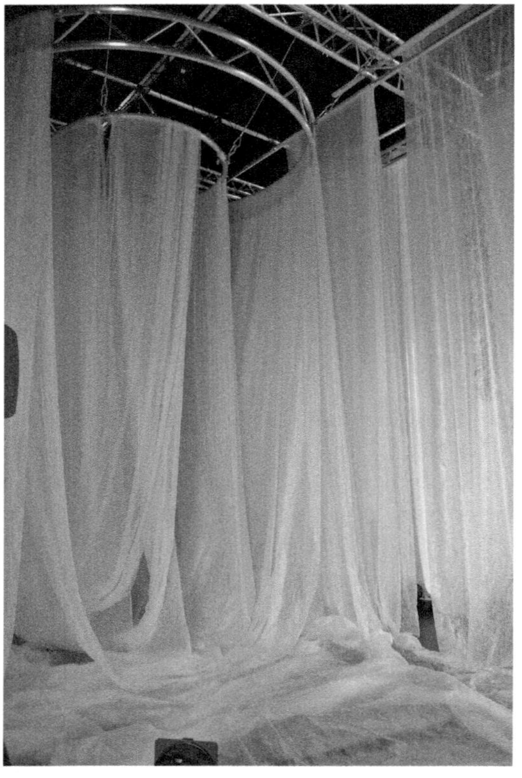

Figure 8.1 View into the setting.

fabric or sat down in the center, spending around 10 minutes inside. Afterwards, they were interviewed about their perception of being able to control the setting and being controlled by it and asked whether they saw any connection between the reactions of the setting and their own behavior. The results of the qualitative interview analyses were published in several papers the researchers presented at various international conferences.

A few months into the project, the researchers also contributed to the production process of a theater performance. The theater director used their setting as a point of departure for the discussion of the chances and risks wearable technologies imply for sustainable, self-organized living strategies. The script was written on the basis of edited excerpts from the first interviews and research results; the public performance was presented in the lab three times to an audience of around 30 people each time.

This brief insight from my ethnographic field research introduces the question that is central to this chapter: What is the role of boundary objects (Star & Griesemer, 1989) in artistic research contexts? If boundary objects, which are jointly developed in collaborative work environments, are understood as means to negotiate different understandings and translate between disciplines, in which way do they influence the emerging research dynamics? My analysis of two university-based artistic research projects—one in the field of digital musical instruments, the other examining computer-controlled atmospheres—revealed different kinds of research dynamics. In the first case, developing a digital musical instrument required discussing design and research practices and negotiating aesthetic as well as bodily preferences regarding its playability to decide about its material and formal characteristics. Thus, the object was in the center of attention. Over the course of its creation, institutional conflicts and stereotypical ideas about "artistic" or "scientific" researchers rigidified rather than being used productively. This led to individualized working patterns and almost hindered collaboration among the researchers rather than enabling it. In comparison, the second case was structured around the broader field of computer-controlled environments. This also required constant discussion of aesthetic and material preferences, different terminological understandings and methods to foster a translation and mutual understanding across several disciplines. These kinds of clarification kept the flow of collaborative research and design processes going, and the researchers managed to integrate their subjective backgrounds and interests into synchronous research dynamics, allowing for collaboration along distinct aesthetic and thematic ideas.

In the following, I will intertwine Star and Griesemer's concept of boundary objects with excerpts from my field research to discuss the emerging research dynamics of the two cases I examined. At the end of this chapter, I will elaborate on the idea of synchrony and how synchronous research dynamics can be fostered in artistic research contexts.

Boundary Objects as Mediators

When Star and Griesemer first published their idea of boundary objects, they defined them as mediators in heterogeneous research groups, as the following quotation shows.

> In conducting collective work, people coming together from different social worlds frequently have the experience of addressing an object that has a different meaning for each of them. Each social world has partial jurisdiction over the resources represented by that object, and mismatches caused by the overlap become problems for negotiation. ... Management of these scientific objects—including construction of them—is conducted by scientists, collectors and administrators only when their work coincides. The objects thus come to form a common boundary between worlds by inhabiting them both simultaneously. ... Intersections place particular demands on representations, and on the integrity of information arising from and being used in more than one world. Because more than one world or set of concerns is using and making the representation, it has to satisfy more than one set of concerns. When participants in the intersecting worlds create representations together, their different commitments and perceptions are resolved into representations—in the sense that a fuzzy image is resolved by a microscope. This resolution does not mean consensus. Rather, representations, or inscriptions, contain at every stage the traces of multiple viewpoints, translations and incomplete battles. (Star & Griesemer, 1989, pp. 412–413)

In this explanation, the authors stress that the collaborative creation of boundary objects does not prevent or solve potential queries or conflicts. On the contrary, its joint construction enables the ability to negotiate different interests, approaches, and meaning-making strategies. With its emergence and evolution, the boundary object is a tangible result of negotiations and discussions, and it even materializes different viewpoints, preferences, and working practices. Consequently, its characteristics serve meaning-making and translation in and across the various disciplines, practices, and subjects that contributed to its creation. This explains why a boundary object allows each of the actors to use it both on a "strongly structured" individual and on a "weakly structured" common level (Star & Griesemer, 1989, p. 393). In the following, I will analyze the role of the digital musical instrument in the corresponding artistic research project and shed light on a boundary object with a reverse structuration.

Back in the Field

The researchers in the second project I examined ethnographically stemmed from seven different disciplines—musicology, computer science, sound engineering, cultural studies, product design, audio communication, and cognitive psychology—and two universities—a technical university and an art university in Germany. They were carrying out research on digital musical instruments as well as on related topics such as the perception of electronic music, the spatialization of sound, and performance aspects related to digital musical instruments. To this end, they developed new musical instruments from scratch. In addition, most of them had experience designing and/or performing with digital musical interfaces; four researchers also had an acoustic musical background.

Three out of the seven researchers—product designer Susan, sound engineer John, and computer scientist Michael, who were all experienced tangible interface developers—had jointly developed the bellow-based digital musical instrument displayed in Figure 8.2 (called "the instrument" hereafter). It consisted of three parts: a Plexiglas box on the bottom, which was tied to the thigh; an upper part, also made from Plexiglas, which was strapped to the hand; and a black latex bellow, which connected these two parts. The hand part included five capacitive sensor surfaces so that each finger could be used to play the instrument. There were two valves and two microphones in the lower part as well as a button to switch between the sound modes. Inside the bellows, there were LEDs illuminating it while playing as well as light sensors, which contributed to the sound-synthesis process. These material and formal qualities of the instrument were developed first, the sonic functions were added at a later stage. Pulling and pushing the bellow up and down produced airy sounds; shaking and bending the bellow quickly sounded as if a helicopter was in the space; and combining the capacitive sensors while simultaneously pushing the bellow down emitted tonal, harmonic sounds. The researchers had not only created the instrument but also performed with it internationally, and the instrument existed in two versions, one for right- and one for left-handed musicians. The researchers sat on a chair while playing the instrument together; each instrument was connected to a notebook computer and a wireless network which served the sound-synthesis process.

I accompanied the researchers of both universities while they designed elements of the instrument in a wood or textile workshop, during team meetings, and during their individual work phases. Over the course of a year, stereotypical ideas about

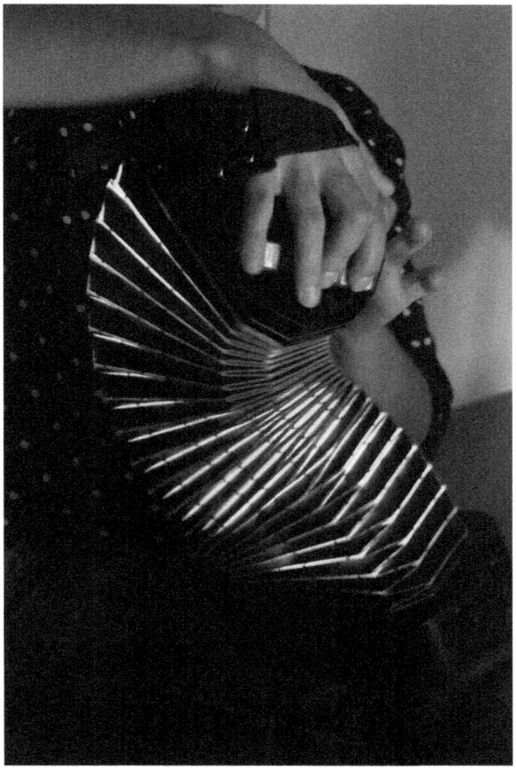

Figure 8.2 Susan playing the instrument.

"artistic" and "scientific" procedures rigidified without being discussed among the researchers, mental barriers increased and hindered collaboration between them instead of enabling it. This divide was also visible in the office spaces of the technical university, where most of the text work and interview analysis was done, and those of the arts university, where sound coding, etching, sanding, and the crafts work involved in constructing the instrument was carried out. While the former were usual office spaces equipped with desktop computers, the latter resembled workshops with mood-boards, soldering tools, and models furnishing the space. The different tasks were split according to the researchers' competences and carried out in parallel lines, thereby hindering the development of a synchronous working rhythm and creating only a few opportunities for, for example, feeding from joint article writing back into the construction and further development of the instrument, and vice versa. In an interview with musicologist Tom who was employed at the technical university, he stated the following:

> Well, the instrument came a bit—somehow in a way—quite as a surprise. I didn't really realize how they came up with it. One day I saw that the [three of them] had done a sort of design workshop. … There was documentation, and then, suddenly, the thing was there and since then, they've been developing it further. …

> Of course, connection points [to our research] may always arise. The problem always is ... if you deal with it—if I may say, academically—the problem is: our approach is always to achieve universality. We look for general tendencies. And when dealing with one very specific artifact, the question always is, what are we able to learn from it that goes beyond it, that can also be found in different instruments? (Tom, interview, 4 Feb. 2015)

On the side of the researchers from the arts university, who developed the instruments, one reaction to the instrument became to be similarly negative. Michael had wanted to achieve a certain level of complexity to increase the instrument's expressivity (Michael, interview, 27 Jan. 2015). At a quite early stage in the research project, he already perceived the relation between complexity and control of the instrument as imbalanced. Ten months into the project, in our second interview, he elaborated on the development process of the instrument in more detail. Michael stated that he had always been interested in exploring a broad range of aspects instead of thoroughly examining one detail, and that this was linked to his interdisciplinary education (Michael, interview, 3 Nov. 2015). He went on to say:

> And the second aspect is connected to the way we approached [the construction of the instrument]. A very *naive* approach to "just quickly sketch such a thing" has led to a *terribly* complicated device, in which I see *many*, *many* dead ends. ... Therefore I'd like to say, "OK, this is one direction in which we went," and now I would like to open up the space and create *at least* one more [instrument] that consciously contrasts the [first] instrument. (Michael, interview, 3 Nov. 2015)

As a consequence of this viewpoint and because the instrument "has become a monster, which has gotten out of control" (Michael, interview, 3 Nov. 2015), Michael abandoned the instrument in the second half of the research project. Instead, he developed and played different instrumental setups that he called "sketches," which "focus on different aspects other than the instrument, but which are *equally* legitimate considering the acoustic-digital approach of hybrid instruments" (ibid.).

One more aspect is noteworthy regarding the reactions toward the instrument. Each time I accompanied Susan, Michael, and John as they rehearsed in the studio, I perceived the atmosphere between them as markedly tense. Even though Star and Griesemer emphasize that they do not have any normative expectations of boundary objects, they stress its material form as carrying traces of "incomplete battles" (1989, pp. 412–413). In the case of the instrument, one could say that instead of solving tensions, it brought them to light. With its emergence, the conflicting opinions among the researchers were unveiled all at once. While the instrument served Susan, John, and Michael as an artifact to make meaning and to structure their research as well as joint work over the course of the project, it had the opposite effect for the researchers from the technical university. Still, to a certain degree, the latter also used the instrument to deduce knowledge and to clarify their notions of digital musical instruments. But since they had not contributed to its construction, the boundary object was even less well structured for them than it was for Michael toward the end of the research project, when he abandoned it to focus on different instrumental setups.

In sum, one could say that the requirements the instrument was supposed to fulfill were too diverse, resulting in a structure hardly adequate for its further use: for some of the researchers, it was too weakly structured to make sense of it on an individual level, for some it was too strongly structured to carry out further research with it. On a common level, it was considered both an object *for* and *of* investigation, almost overcharged with material, formal, and functional qualities resulting in an overly complex structure. One could say that it therefore failed to mediate and translate between the different disciplines and motivations involved. Rather it contributed to individualized working patterns.

Aesthetic Framing of Boundary Objects

In a discussion on the concept of boundary objects in artistic research settings, Borgdorff develops the following argument. He states that the material and practices used for the development of a boundary object, the environment in which it is created, and the questions it addresses define a boundary object's character (Borgdorff, 2012, p. 119). The artifact as well as the knowledge it evokes can therefore be considered as post- or transdisciplinary: boundary objects in artistic research contexts are not necessarily artworks; neither are they mere representations of results or processes. They are created with the help of artistic methods and can therefore be considered as art to a certain degree. In this manner, boundary objects enable meaning-making both in academia and in artistic environments (ibid., pp. 119–121).

This implies that boundary objects do not only influence the work processes of artistic research settings. In addition, they are often also used to disseminate research results, for example in visual displays, in theater, dance, or music performances, and in sound installations. Therefore, they address both a sensuous and a textual level of perception. In addition to artistic presentation formats, boundary objects are used to make research processes transparent. Practice-based artistic research master's or doctoral programs, for example, usually require a written reflection on the final result. This serves to outline the research methods used as well as a broader theoretical contextualization of the practical, artistic part (see e.g. Ambrožič & Vettese, 2013; Leiderstam, 2009; Wilson & Van Ruiten, 2014). In exhibition settings, research processes are often presented in the form of preliminary studies, drawings, or mappings, which accompany the actual artwork or result. Hence, boundary objects that served as a method during the work process can become a means of interpretation and analysis in staged environments. Bippus discusses the potential of modifiable laboratory installations as "critical knowledge *productions*" (2012, pp. 118–119; original emphasis). As opposed to final artworks, which are to be interpreted, such modular settings artistically stage socially relevant topics. In this manner, they reach beyond the usual public of art experts and stimulate lay people to debate and reflect upon these questions as well (ibid., p. 119). Modular settings or installations that combine artistic elements with knowledge-related questions therefore facilitate reflective processes (ibid.).

The preceding paragraphs show that boundary objects in artistic research settings have diverse forms, inhabit varying roles throughout the research process, and address several types of recipients. In the field of artistic research, the materiality of boundary objects opens up different analytical perspectives. This aspect is relevant already during the research process—that is, researchers contribute to the development of boundary objects with their knowledge and expertise. Since they create the artifact during the research process, it influences the research setting itself, by altering the procedures, potentially requiring changes in the methods used, and raising new questions. In this manner, boundary objects also impact the types of results, interpretations, knowledge, and presentation formats developed during artistic research projects.

Against this background, interpreting the computer-controlled environment I briefly described in the very beginning of this chapter as a boundary object is unambiguous: the setting served to mediate across the different disciplinary backgrounds involved in its development and in the theater performance—acting, dramaturgy, media art, cultural studies etc. In Star and Griesemer's words:

> Boundary objects are objects which are both plastic enough to adapt to local needs and the constraints of the several parties employing them, yet robust enough to maintain a common identity across sites. ... They have different meanings in different social worlds but their structure is common enough to more than one world to make them recognizable, a means of translation. The creation and management of boundary objects is a key process in developing and maintaining coherence across intersecting social worlds. (Star and Griesemer 1989, p. 393)

The theater performance added yet a further layer: in its specific aesthetic framing, the setting's functionality was artistically staged to communicate socially relevant ideas to an audience. Through this structural and interpretative enrichment, that is, through simultaneously being a research tool and an aesthetically perceivable entity, the performance served as a means of communication between an enclosed research community and a broader audience. It offered a sensuous experience of research processes and methods, through which the audience gained personalized knowledge. The relation to the broader theoretical context of wearable technologies and sustainable self-organized living furthermore enabled the audience to acquire generalizable knowledge. As a tangible yet text-based entity one could aesthetically experience and thereby comprehend through different senses, the performance enabled subjective sensuous approaches to the topic and offered overarching socially relevant ideas at the same time.

Summed up, the structuring role and influence of the setting as a boundary object was obvious throughout different stages of the research project. Through jointly developing an artistically staged responsive environment, Anna, Nick, and Steve got to know each other's interests in the setting and acknowledged their distinct contributions to its development. They could approach it from their individual disciplinary backgrounds while simultaneously using it as a common point of departure for the readjustment of their interview questions or the further technical and aesthetic refinement of the setting itself. Similarly, the meaning they derived from the setting was both personalized or related to their individual disciplines and universally valid, so that they could put forward hypotheses about the effect of computer-controlled environments on the perception of affective atmospheres both in publications and for further funding applications. Finally, their discussions about aesthetic aspects as well as research strategies always resulted in a joint decision about the further procedure. In the design process of the boundary object, the researchers developed what I would like to call a synchronous research rhythm.

Synchrony

To examine how synchrony emerges between several people, it is insightful to look at the notion's origins in psychology and psychotherapy. Koole and Tschacher provide the following explanation for synchronization between people:

> When interaction partners become synchronized, they become adapted to each other's rhythms and cycles of activity, like people who are dancing together. This mutual adaptation may mean that interaction partners come to display similar behaviors. However, interpersonal synchrony does not always involve imitation or mimicry. ... For instance, if one interaction partner nods her head in response to another's hand movements, this still qualifies as interpersonal synchrony. Interpersonal synchrony thus depends on the mutual timing of responses, regardless of the precise form of these responses. (Koole and Tschacher, 2016, p. 9)

In relation to the patient–therapist relationship, the authors state that synchrony indicates increased mutual comprehension, which is mirrored in bodies that behave, move, and perceive similarly (Koole and Tschacher, 2016, p. 9). In line with that argument, Wheatley, Kang, Parkinson, and Looser claim that synchrony in groups increases the feeling of empathy between the individuals as well as their attachment and belonging to that group (2012, pp. 594–596). They state: "The ability to synchronize with others enables emotion contagion. ... [It] allows us to spontaneously embody the affective experiences of familiar others, and this embodiment fosters better social understanding" (ibid., p. 594). Furthermore, in reference to Polanyi (1983) and his exploration of implicit forms of knowledge, Gill claims that bodily synchronization is central to the transfer of tacit knowledge and hence to the joint engagement in practices as well as to the negotiation and "formation of a common ground" in the context of distinct points of view (2007, pp. 569, 600).[2]

Similarity of behavior does not imply, however, sameness. Rather, synchrony refers to activities that are carried out at the same time, that are coordinated, composed, and that follow a common rhythm. Wheatley and colleagues argue that "coordination of distinct roles benefits from an understanding of how the system moves as a whole," whereas the coordination of similar activities evokes the feeling of representing one and the same system, of enacting the same roles, of resemblance (2012, pp. 595–597). This is a relevant aspect for understanding group dynamics, because the feeling of belonging and resemblance can have both negative and positive consequences. On the one hand, strong social bonds lead to enhanced feelings of security, pleasure, and enjoyment. The latter are linked to the bodily reward system. Semin and Cacioppo define synchrony "as jointly and simultaneously recruited sensory motor processes that are evident in a neurophysiological mirroring of the producer by the perceiver" (2008, p. 123). Since these processes of neurophysiological mirroring take place automatically, increased social ties and mutual understanding can be achieved efficiently, without needing to invest much effort. Therefore, synchrony is perceived as specifically rewarding and can even be considered as desirable as far as fulfilling basic needs of belonging (see also Wheatley, Kang, Parkinson, & Looser, 2012, p. 596).

On the other hand, Wheatley and colleagues explain that "[brain] areas involved in the perception and production of dynamics literally entrain to the same oscillatory rhythm. ... Group dynamics may serve to overcome individual rhythms, thereby providing a more compelling beat to which to entrain" (2012, pp. 596, 599). Therefore, if group dynamics have a muting effect on individuality, synchrony might likewise lead to a decreased self-perception (ibid., p. 597). Consequently, the authors state that even though synchrony automatically happens, attempts at deliberate counter-movements—what they describe as "engineering disconnection" (ibid., p. 600)—might be fruitful for group dynamics. That is, instead of muting or overcoming differences, it might be more productive to be aware of them and to consciously integrate them into research processes.

In sum, getting in the *same* mood and developing the *same* tempo and the *same* rhythm is neither necessary nor desirable in research settings. Rather, one can say that just like different opinions, various work patterns and practices can exist in synchrony, that is, in correspondence with each other. It is a matter of coordinating differences, not of eliminating them. Even though these differences might merge into a common rhythmic figure—literally, in the case of synchronized instruments, or figuratively in the case of a work routine—they escape deliberate organization. While synchrony automatically occurs via neurophysiological oscillations that happen on the same rhythm (Wheatley, Kang, Parkinson, & Looser, 2012, p. 596), the pure ability or will to synchronize does not necessarily lead to synchrony.

Synchronizing Practices

What implications do these elaborations have for research settings in which aesthetic practices and mediation between various disciplinary backgrounds are crucial? Going through the questions of synchronization applied to both case studies results in the following considerations.

In the case of the instrument, the team meetings with all researchers from both universities I participated in had an agenda but no schedule. Neither did they have a limited timeframe, nor did they include breaks. The arts university researchers' internal team meetings proceeded similarly: lunch or coffee breaks were carried out "according to demand," and if necessary, the meetings would continue after these breaks. The technical university researchers' internal meetings usually had a predetermined start and end time. They were restricted to one and a half or two hours maximum, usually followed by a joint lunch or coffee break. These meetings mostly served a structured overview of upcoming publication or conference deadlines, an update on the individual work packages and questions or challenges to resolve. In some of the meetings I observed, the difference between "experimental" or "artistic" and "scientific" methods the researchers used to investigate the research questions led to tense, conflicted discussions and fostered stereotypical ideas of each other. Rarely did they discuss a topic, specific concept, or text to clarify their thematic interests and notions underlying each of their research topics. The researchers' individual competences were split in correspondence with the work packages and in an efficient manner so that both the instrument and the publications were well received in the project's artistic and academic peer groups. Yet, the researchers' perceptions of their seemingly different approaches oftentimes hindered the establishing of a productive work environment, which finally resulted in individualized rather than collaborative work patterns. Overall, despite the fact that the researchers' activities were loosely coordinated in one and the same research project, conflict situations that arose due to individual sensitivities and resentments stood in stark contrast to the possibility of getting along well with each other and finding a common, synchronous working rhythm.

In the case of the setting, team meetings did not seem to be structured at all. One of the three researchers would ask their fellow colleagues whether they should get together to discuss the setting's further development, the performance, or an article to be submitted. According to each person's availability, the meeting started immediately, half an hour, or two hours later. Phone calls, other project partners passing by, or coffee and lunch breaks quite frequently interrupted the meetings. These "forced breaks" did not result in negative affects such as anger or impatience. On the contrary, my impression was that Anna, Steve, and Nick had become used to these interruptions and adapted their behavior accordingly. This implied that the three had established a sort of intuitive rhythm of gathering, parting, and coming back together for their joint work. Steve and Anna automatically worked on their own projects when Nick followed his daily rhythm, which included a short siesta after lunch followed by a coffee and a small piece of chocolate to get him back into working mode. During this time, Steve and Anna did not work in their lab spaces because Nick usually used their shared sofa for his siestas.

The most important aspect in hindsight to these team meetings is that they regularly opened up a space to discuss the theoretical framing of the setting through clarifying notions such as atmosphere or affordances (Gibson, 1986) and to jointly work on the refinement of the setting, thereby commonly deciding about its material and formal characteristics. Such an openness and space for discussion was rare in the case of the instrument.

Implications for the Field

What conclusions can we draw from the different research dynamics emerging around the boundary objects in both cases? Which implications do these differences have on the organization of research processes and on the development of a research environment that fosters synchronous research patterns rather than to prevent them?

The emerging dynamics and their development in artistic research contexts depend on a specific constellation of people, experiences, material, spatial settings, timing, and institutional contexts. My field research revealed that an object, which requires all of the researchers' attention, affords a different kind of organization to promote mutual understanding between the different people involved in its construction than a topic or field of investigation. Objects are strongly structured via their concrete material, formal, and functional characteristics. These characteristics quickly lead to decisions about whether or not one can make use of the object. The case of the instrument revealed that an object can be stretched to a certain degree only, since overcharging it with functions can result in its being too complex. This, in turn, risks an object's collapse or being rejected because it can fulfill a certain number of requirements only. In comparison, fields, concepts, or topics that are more loosely structured on a common level are more flexible. They serve as a means to negotiate different understandings between disciplines and in a second step deciding on whether and in which way they can be applied. These discussions better serve to foster what could be termed interdisciplinary competence: they increase mutual understanding between disciplinary backgrounds as well as different practices, which is essential in synchronization processes.

Against this background, a first practical conclusion is the following: given that timing, a certain mood, flow with materials, and the spatial setting are so essential to artistic practices, these aspects should be considered in the development of artistic research projects. Practically speaking, this includes considering the arrangement of office spaces and workshops and instituting flexible office hours and space usage to accommodate individual working patterns, alongside scheduling regular meetings to create sufficient room for exchange between the researchers involved, and thus enabling the synchronizing of various distinct research dynamics. Depending on the size of the team, this might even require employing a project coordinator who takes care of administrative and organizational tasks.

From reconsidering the development of a research setting, re-examining evaluation criteria is just a logical consequence. The fact that research settings are constructed in accordance with field-inherent logics and approaches and that these approaches are mirrored in the research dynamics and outcomes is not surprising as such. However, if these logics continue to serve as evaluative criteria for the generated results, research projects will persist in being caught in a circle of self-fulfilling prophecy and constant legitimization. The criteria currently used to evaluate artistic research projects are similar to those used to evaluate any other academic research project. They focus, for example, on a quantitative evaluation of publication indices, acquisition of further third-party funding, and the number of PhD students supervised during the project. It is questionable, however, whether funding bodies and the participating researchers should continue to evaluate artistic research projects on the basis of their results and research output. Rather, the projects' emerging research dynamics and the idea of synchrony could serve as alternative or additional indicators of a successful and productive research project, as these factors account for each researcher's individual approaches while at the same time analyzing their integration and exchange with further methods and ideas. Such an evaluation would be qualitative, and possibly be based on the subjective experiences and individual feedback of the researchers involved. In practical terms, this could be realized within the team: for example, in one or two brief annual meetings with a project coordinator or the principal investigator. Continual reflection on both the development of team dynamics and research progress is essential to remaining aware of and incorporating each researcher's aims, experience, and motivation.

Such an analytic approach is in accordance with the overall dynamics of the field of artistic research. Negotiations of different perspectives and discussions about conflicting standpoints should be enabled rather than being muted. In line with this argument, one could look at how and whether differences can be integrated in a productive way, to account for the field's diversity in methods, topics, and theories.

Against this background, I would like to raise another question. What role do institutional boundaries, politics, and missions play in the development of common research rhythms? If artistic research projects are established on temporary grounds only, in what way can institutional structures be used for these projects without adhering to the implied politics of the institutions? Maybe artistic research practices can serve to push institutional boundaries and limits. To come back to the notion of synchrony: instead of imitating and adapting to the rhythm indicated by institutional politics, artistic research practices could serve as a counter-movement, an "engineered disconnection" (Wheatley, Kang, Parkinson, & Looser, 2012, p. 600) that sheds light on how the institution functions as a whole. In this way, research projects and their individual researchers from artistic, scientific, cultural, and other disciplinary backgrounds could manifest their subjectivity and distinct working patterns as a temporal part of that institution and as such raise awareness regarding how the hosting institution works.

However, in opposition to this idea of synchrony as being able to be directly and deliberately influenced, Koole and Tschacher state: "The emergence of synchronous behavior does not depend on intentions or any other quality of the individuals who are behaving in synchrony. Rather, synchrony arises as a self-organized behavioral pattern from people's mutual interactions" (2016, p. 5). Therefore, if one is to understand artistic research as a tool for, among other things, understanding institutional settings, an environment that allows for the *integration* of individual research patterns, motivations, and practices into synchronous working rhythms similarly needs to allow for a *decoupling* and rearrangement of its elements, for example, in a situation where the researchers are simply unable to synchronize. Such disintegration, however, should avoid instability or precarity. Rather, both the institution and the researchers should be able to persist in their individuality, join other collectives, and proceed with their ideas and working modes; that is to say, to conclude their research endeavors.

Notes

1 This chapter is based on edited excerpts from the results of a detailed ethnographic long-term study first published in Schindler, 2018.
2 Based on an ethnographic analysis of a further musical instrument design project, Schindler (2015) differentiates between various forms of tacit knowledge and their transferability between researchers from several disciplines.

References

Ambrožič, M., & Vettese, A. (2013). *Art as a Thinking Process: Visual Forms of Knowledge Production*. Berlin: Sternberg Press.

Bippus, E. (2012). Modellieren ästhetischer Wissensproduktion in Laboratorien der Kunst. In M. Tröndle & J. Warmers (Eds.), *Kunstforschung als ästhetische Wissenschaft: Beiträge zur transdisziplinären Hybridisierung von Wissenschaft und Kunst* (pp. 107–126). Bielefeld: Transcript.

Borgdorff, H. (2012). Boundary Work: Henk Borgdorff interviewed by Michael Schwab. In F. Dombois, U. M. Bauer, C. Mareis, & M. Schwab (Eds.), *Intellectual Birdhouse: Artistic Practice as Research* (pp. 117–123). London: Koenig Books.

Gibson, J. J. (1986). *The Ecological Approach to Visual Perception*. New York: Taylor & Francis.

Gill, S. P. (2007). Entrainment and Musicality in the Human System Interface. *AI and Society: Knowledge, Culture and Communication, 32*, 567–605.

Koole, S. L., & Tschacher, W. (2016). Synchrony in Psychotherapy: A Review and an Integrative Framework for the Therapeutic Alliance. *Frontiers in Psychology, 7*, Article 862.

Leiderstam, M. (2009). See and Seen: Seeing Landscape through Artistic Practice. *Art and Research, 2*(2), 1–8.

Polanyi, M. (1983). *The Tacit Dimension*. Gloucester: Peter Smith.

Schindler, J. (2015). Expertise and Tacit Knowledge in Artistic and Design Processes: Results of an Ethnographic Study. *Journal of Research Practice, 11*(2), Article M6. Retrieved from http://jrp.icaap.org/index.php/jrp/article/view/494/421.

Schindler, J. (2018). *Subjectivity and Synchrony in Artistic Research: Ethnographic Insights*. Bielefeld: Transcript.

Semin, G. R., & Cacioppo, J. T. (2008). Grounding Social Cognition. In G. R. Semin & E. R. Smith (Eds.), *Embodied Grounding: Social, Cognitive, Affective, and Neuroscientific Approaches* (pp. 119–147). Cambridge and New York: Cambridge University Press.

Star, S. L., & Griesemer, J. R. (1989). Institutional Ecology, "Translations" and Boundary Objects: Amateurs and Professionals in Berkeley's Museum of Vertebrate Zoology, 1907–39. *Social Studies of Science, 19*(3), 387–420.

Wheatley, T., Kang, O., Parkinson, C., & Looser, C. E. (2012). From Mind Perception to Mental Connection: Synchrony as a Mechanism for Social Understanding. *Social and Personality Psychology Compass, 6*(8), 589–606.

Wilson, M., & Van Ruiten, S. (Eds.). (2014). *SHARE Handbook for Artistic Research Education*. Retrieved from https://www.elia-artschools.org/userfiles/Image/customimages/products/120/share-handbook-for-artistic-research-education-high-definition.pdf.

9 Figurations of Hybrid Ecologies in Artistic Practice

Desiree Förster

> We imagine a future that is completely dark
>
> *BIO[lum]SKIN*

What does lived experience mean in times of environmental crisis? The first-person perspective of the lived body, which in phenomenology is foundational to sensual perception and knowledge creation, seems to be unable to grasp processes on the planetary scale such as climate change. The environmental crisis is so extensive and neither temporally nor spatially understandable to an individual that some scholars (such as Carlson, 2002; Berleant, 1995) call for a new environmental aesthetic, which would enable an appreciation of nature from a new point of view. Such a re-situating of human agency into its natural environment is being discussed in new approaches to materialism as well. New Materialism, as it developed as a theoretical field in the humanities since the late 1990s, joins several fields in their common way of re-thinking subjectivity in emphasizing the role non-human powers and processes play on various levels of life and sense-making. In this text I will explore how new aesthetic practices on the intersection of art and design develop forms of incorporating non-human agencies into the lived and sensual experience or expand the human body towards its animated, vital environment. Using concepts from phenomenology, New Materialism, and actor–network theory, I will discuss these practices as examples for a new aesthetic, which is not so much an environmental aesthetic but already goes beyond the systemic approach associated with it.

BIO[lum]SKIN: An Aesthetic of Interrelations

We see the naked back of a human body; the face is wrapped in a material that looks like dried skin or vellum (Plate 7). It covers the face like a veil and runs down from the back of the head, along the neck, spreads out in spine-like structures around the shoulders that continue all the way down. The human figure almost looks like a dragon or a relic from the Stone Age – a hybrid coming to live from imaginary tales or ancient times. What we see, instead, is a speculation of a future in which the human itself is moved into the realm of fiction, a future where symbiosis and hybridization replace the individual organism. In this imaginary titled *BIO[lum]SKIN*, developed by students of the Institute for Advanced Architecture of Catalonia (IaaC) in 2016, the environmental crisis associated with climate change, pollution of air and oceans, has led to an ecological catastrophe that forced the surviving species to a radical shift in the ways of becoming: to survive is to interconnect. Humans, we learn from the text that accompanies the pictures on the website,[1] have undergone a metamorphosis and entered a symbiotic relationship with sea creatures. "The organ", which runs from the head all the way down the back of the body, has grown to provide a habitat for these organisms that would otherwise die.

The project speculates about a future that considers the changing oxygen levels in the world's oceans, as we record today. Since warming ocean waters cannot hold as much oxygen as colder water, the drop of oxygen levels leads to a change in the ocean's species diversity. The group presents the human species as an altered state of organisms, in symbiotic or parasitic relationship with sea creatures, which lost their natural habitat due to a lack of oxygen. Therefore, the human develops a new organ that hosts these creatures and provides oxygen responsive skin.

"The future is dark", as the artist group states in the project description. It is a future in which those who relate survive, those who can share the remaining resources and find ways of co-producing and co-evolving with others – that are not necessarily human. If relations, relating to others, is the new key for flourishing and survival in apocalyptic times, then art can be the place that allows attunement to otherness, experimentation with ways of engagement, and thereby create novel sensory-emotional values. This aesthetic reminds us of the concept of "relational aesthetic", as developed by curator Nicholas Bourriaud in the 1990s (Bourriaud, 2009), to describe artworks based on human relations and their social context. Works like *BIO[lum]SKIN* expand towards non-human relations, however, and contain biocultural processes and how they embed bodies into environments that are shared with other entities, such as chemicals, bacteria, and non-biological materials. These others are not perceived in our everyday life, but they gain meaning when they become perceivable as the very elements that structure our being: oxygen is taken in through breathing, an activity we constantly perform and yet are not aware of most of the time.

Oxygen enabled life on earth to develop in the first place, and its concentration in the atmosphere defines which organisms can survive and which cannot. A change of the atmospheric condition as imagined in *BIO[lum]SKIN* might at first affect the sea creatures – but the hydrosphere, the water component of the Earth, is not separate from the biosphere, the habitat of human beings. All the Earth's spheres are in constant exchange of matter and energy. A compositional change in one of the spheres consequently effects all the other spheres as well. In *BIO[lum]SKIN*, it is the materiality of the human body, its potency, that serves to compensate for such a change.

If the body and its bio-chemical structure becomes central for the life-sustaining relations in the environmental crisis, what role does consciousness and lived experience play? The discussion about the significance of the human mind in relation to its being in the world, which has played a role in many knowledge disciplines since Descartes' *Meditations* (1641; first English translation, 1647), is yet again being revisited in the light of the global crisis, often associated with the Anthropocene. Questioning the role of consciousness and lived experience challenges the concept of the human subject as superior to non-human creatures. With the Anthropocene we enter a new geological age characterized by the human impact on Earth. Besides its emphasis on the human as a planetary force, the Anthropocene as a concept of thought urges attention towards the living and non-living elements that inhabit the Earth's spheres. This focus on the underlying composition of our local and global environments leads to a shift in thinking the human: although naming the current geological age after the human species might indicate the central impact it has on the planet – but with a closer look it becomes clear that this impact does not speak so much of human superiority but more about the myriad of ways in which human organisms and the Earth's spheres are interconnected. The Anthropocene is not the age of the human, it is the age of mutual interrelations across species and material boundaries, complex material cycles and endless recurrence of newly induced materials such as plastics, radioactive waves, carbon and garbage. But awareness is not necessarily followed by change. The scope of these connections is not to be understood from the perspective that puts the human subject at its center. But how to tackle the environmental crisis characteristic of the Anthropocene instead? If we cannot grasp global warming, can we still act on it? The environmental philosopher Timothy Morton suggested the concept of the "hyperobject" to refer to things on scales that are beyond human comprehension. Taking the existence and impact of such object into account, does not mean that we regain intimacy with and knowledge of the world. The opposite seems the case: as Morton states, planetary awareness does not increase the understanding that we are the world – rather it makes apparent that we are not (Morton, 2013). However, we can examine how we come to our conceptions of being in the world and open ourselves to those processes that are involved in generating them in the first place.

Human Subjectivity as Energy-in-Transition

Art projects such as *BIO[lum]SKIN* imagine the human as being in constant exchange with its surroundings and not in opposition to it. In face of the environmental crisis, a new post-human agency is proposed, in which the human subject is not opposite to the world, is not superior because of its higher-conscious abilities such as self-awareness. What matters instead is the capability of the sensitive body to connect to the world on multiple levels. Agency in post-apocalyptic times emerges out of symbiotic relationships with non-human others. What is proposed in the speculative work is thus not the constitution of a subject-as-a-whole but of nature-as-a-whole, in which the human participates because it is already embedded into the world via metabolic pathways. *BIO[lum]SKIN* proposes a syn-aesthetic experience, one that reveals the being as being part of a whole, a world that is shared.

The environmental crisis comes with a disruption of human self-understanding. However, in order to move from the estrangement that Morton refers to, towards a perspective that is able to integrate this estrangement into a new whole, it is necessary to rethink which agents matter in this new projection. Such a "new whole" would not regard humans as opposites or at the mercy of nature, but as connected to it in ways that are not completely rationalizable and therefore necessarily strange. The estrangement from nature diagnosed by Morton is thus reinterpreted as a productive force within the discourse as it is currently conducted by other contemporary philosophers, such as Isabelle Stengers and Bruno Latour. One possible figure to think nature as a whole without ignoring the strange and uncontrollable as being part of it, is proposed by Bruno Latour. His concept of Gaia, a new figure for the Earth, is a "name proposed for all the intermingled and unpredictable consequences of the agents, each of which is pursuing its own interest" (Latour, 2017, p. 142).

Gaia is especially not a figure of unification, it does not offer a global perspective. Instead, Gaia may "reterrestrialize" (Latour, 2017) our existence in emphasizing a partial view on the matters of the Earth instead of reinforcing a global one. Gaia refuses representation and objectification and therefore challenges us to turn towards those materials, organisms and processes that play central roles in the mattering of

Gaia. Gaia therefore is described by Latour as "an animated world, an Earth that vibrates underfoot, no recognizable landscape, no affirmed authority, frightful mixtures, a proliferation of hybrids, scattered members of sciences, industries, and technologies" (ibid., p. 149). In offering aesthetic experiences of how distinct entities start to matter if we take on a such a perspective on the world, artistic projects can offer exactly that partial view needed to move away from the Earth as a representable whole. The whole instead emerges from its parts, which are not objectifiable either, but continues processes composed of diverse agents.

If we embrace the idea of the human being enabled by and emerging from, among others, its biochemical relations with other entities that embed it in a shared environment, we face yet another challenge. If the human being cannot be thought of as distinct from its biological habitat, in what way can we still speak of human agency at all? The anthropologist Samantha Frost has addressed this question in a recent publication. Here, she proposes "bio-cultural creature" as a new term for human beings, while stressing that the emphasis on the biological impact on the formation of the human subject does not mean that it is reduced to a passive emergence from already constituted biological – or cultural – networks. Frost writes in line with scholars such as Judith Butler who understand the material, sexed body as a discursive formation. But Frost moves a step further arguing that the human subject is not only shaped by its material-cultural surroundings but also by its biochemical activities. Being aware that rethinking the human subject in terms of the materialization of historically specific norms (Frost, 2016, p. 121) can lead to an overemphasis of the social context a subject develops in, she tries to define the human subject as co-constructed by environmental structures, without making it all dependent on it. Frost's biocultural creature is characterized as "energy-in-transition", meaning that "the living body's biochemical activities depend on and effect transitions in energy, transitions that themselves are profoundly dependent on a living organism's immersion in and engagement with its habitat" (ibid., p. 119). To not end with a description of the living organism as a mere epiphenomenon of its environment that would re-inscribe yet other dichotomies, she introduces the principle of "activity-in-response", that obtains the dimension of a carried and corporealized history:

> [The] living history is the organismic material and activity through which the current environment has its effects. The proteins and other biomolecules that an organism's cells use to process and respond to chemical signals generated through the organism's encounter with the environment – those proteins and biomolecules are produced, in part, by the responses of previous generations to their environments. As a consequence, the effects that an environment can have on a porous organism are constrained by the responses of those previous generations to previous environments. Similarly, and from the obverse perspective, the responses an organism can muster in response to the provocations of its habitat are constrained by the ways that prior generations' responses shape its own processes of composing and decomposing. (Frost, 2016, p. 123)

For Frost, these constraints follow from a principle of non-contemporeneity that characterizes the biocultural being and its habitats. She identifies a genuine latency between the effects of an environment and the response of an organism. Furthermore, the scope of possible effects and potential response is not infinite – it is structured by prior generations and their interrelations with the environment. Such constrains therefor allow for freedom and novelty to arise, to not reduce the biocultural being to mere response-mechanisms.

The relations between the human being and its surroundings seem to be open for continuous exchange and yet closed, in the way pre-existent patterns of perception, reflection, and behavior structure the exchanges. *BIO[lum]SKIN* offers a figuration that acts out such a dialectic on the scale of the atmospheric surrounding: the changing oxygen levels force sea creatures to leave the ocean and to form a symbiotic relationship with humans. The human body becomes a nourishing habitat for them. Together they form a hybrid that does not dissolve all creaturely differences but adds zones of interpenetration into an existing milieu. The symbiotic relationships that the human being is already in (for example, as demonstrated in recent microbiome research[2]) are made visible and transferred onto a larger scale, connecting the human with a different, formerly separate sphere, the hydrosphere.

BIO[lum]SKIN re-figures the human in an altered state, which projects those relationships with the environment that matter in times of crisis. The re-composition of the human organism brings to the foreground non-human agents and metabolic processes, which matter in a post-apocalyptic scenario. The skin is not a boundary of bodies anymore but instead an organ that extends the organism towards others. Through this extension towards and incorporation of other agencies, creatures that usually do not come to our awareness, start to matter. The sensual features of the wearable offer a multisensorial experience, whereby the imagination of a new hybrid whole is also reflected in the sensory experience (Plate 8). *BIO[lum]SKIN* ultimately imagines the human not as a mere user of natural phenomena but as interconnected with them. The body as composed out of different entities illustrates how things, people and environments co-constitute each other.

Entangle Yourself: *H.O.R.T.U.S.*

Besides the motif of the hybrid as re-imagined in *BIO[lum]SKIN*, the image of the garden seems fruitful to think anew about how intertwined living processes are. While the hybrid still forms a bodily object – although with extended and altered bodily boundaries – the motif of the garden introduces a field in which nature-cultural forces bundle, enabling growth and distribution. The next art work I want to introduce uses the motif of the garden to create a responsive landscape that tackles the relations and feedback systems human beings are embedded in. Not just illustrating these relations, it adds possibilities for communication, acting and learning into a spatial structure that contains living and nonliving elements. Thereby a space is being created that offers different ways of acting and interacting with vegetal agents.

H.O.R.T.U.S. is a project by the design collective ecoLogicStudio and was exhibited first in 2012 at the Architectural Association School of Architecture, London. It "engages the notions of urban renewable energy and agriculture through a new gardening prototype"[3] in using atmospheric media – air as well as computational technology – to interrelate human visitors with algae. I understand atmospheric media here as both natural and computational elements that form the background of our experience.

The term "atmosphere" is polysemantic: atmospheres can be felt as a mood or tone. A work space can have a "good atmosphere"; the atmosphere of a dinner party can be vibrant. Atmospheres can be subject to science, such as in meteorology. Here atmospheres serve the purpose of determining climatic conditions and making weather predictions. In the latter understanding, atmospheric media can be water, light or air. In the former understanding, they are composed of all parameters that affect us emotionally. This includes of course the "other" atmospheric media: the colorful light of a sunset can have a deep impact on how we feel. Likewise, we are impacted by customized adds when we browse the internet. Forming the background of our experience, atmospheres are most of the time not subject to our reflection in daily life. Yet, at times they come to the foreground and reveal how our sensual experience is structured.

In *H.O.R.T.U.S.*, algae contained in plastic bags hang from the ceiling of a space, at the eye-level of the visitor (Plate 9). Connected to the bags are tubes that the audience is invited to breathe into and thereby to feed the algae carbon dioxide. Fed with carbon dioxide, the algae produce oxygen and thereby alter the air in the space. Additionally, these effects are mediated by a virtual layer: via smartphones and terminals inside the exhibition additional information on a particular algae is given, about the oxygen levels inside the bag and how much the algae has grown. The evolvement and status of each algae could also be followed up online. The co-existence and co-evolvement of biological and computational systems plays a central role here: sensing the change in air quality, monitoring live processes, and visualizing feedback loops, becomes an intrinsic part of the spatial arrangement and how being in the space is experienced: breathing the air with others, being a breathing body in a space whose air quality changes due to this very act of breathing.

The breathing body displays the doubleness of the body as described by phenomenologist Maurice Merleau-Ponty. The body is both visible as object of experience and invisible as the subject of experience. The body intertwines the sentient and the sensible and is therefore not an object, but an "pre-objective unity" (Merleau-Ponty, 1968, p. 141). This duality for Merleau-Ponty permeates all being. And perceiving means exploring the invisible depths of each quality. In *H.O.R.T.U.S.* the invisible quality of our breath is being explored. The relationship between humans and algae facilitated by atmospheric media opens up smallest, invisible levels of engagement enabled by the energetic and material flows that embed living organisms in their environment.

Breathing Together

In *H.O.R.T.U.S.* what usually lies in the background of our experience – the being embedded in environments through breathing, exchanging air with others – comes to the foreground. This coming to the foreground of processes discloses new meaningful patterns, that can be reflected and acted upon:

> Augmenting our sensory palette through exposure to extrasensory projections opens opportunities for interfacing with imperceptible phenomena. Rendering imperceptible phenomena is most effective through an intermittent or continuous relationship, in which the translation of invisible information is learned over time. Responsive projects that *elucidate* bring clarity to ordinarily unseen and invisible phenomena through methods of visualization and actuation. (Bradley and Holzman, 2015, p. 54)

Projects such as *H.O.R.T.U.S.* create a space that allows for the experience of atmospheric and material surroundings, whereby the human body becomes one bundle within a field of forces amongst others. If human subjects emerge here as part of dynamic interrelations with their surroundings on bodily, atmospheric, social levels, which do not stem from a substantial core, we need different and multifaceted perspectives on the ways human subjectivity plays out that articulate these different layers. These perspectives cannot be left to one discipline alone but must come from diverse practices, historical and cultural backgrounds.

In this line of thought, anthropologist Donna Haraway pleads for new narratives that do not center on the human as the main force and decision maker but as part of a world that is at risk. Global catastrophes like climate change, pollution, species extinction, have not only far-reaching impacts on the human habitat – they also bring to light how the effects do not affect everyone equally. Deep inequalities between cultures, social structures and species emerge that are not only a result of centuries of human inhabitation on earth but a consequence of discriminating systems. At the same time, these inequalities and the distributed attention that is given to the growing ecological crisis, might be yet another reason for its very escalation. Donna Haraway therefor introduces the term "Chthulucene" to not reduce the discussions of the present crisis to the human-centered view of the Anthropocene debate:

> Specifically, unlike either the Anthropocene or the Capitalocene, the Chthulucene is made up of ongoing multispecies stories and practices of becoming-with in times that remain at stake, in precarious times, in which the world is not finished and the sky has not fallen – yet. We are at stake to each other. Unlike the dominant dramas of Anthropocene and Capitalocene discourse, human beings are not the only important actors in the Chthulucene, with all other beings able simply to react. The order is reknitted: human beings are with and of the Earth, and the biotic and abiotic powers of this Earth are the main story. (Haraway, 2016, pp. 11–12)

To narrate these stories, diverse figures, elements and characters are needed, that strengthen ideas of flourishing, of growth: life-affirming stories, that bring life to experience as diverse, in flux, not clean but constantly interrelating with others. In *When Species Meet*, Haraway introduces the concept of "response-ability", as an ethical response towards entangled subjectivities – meaning entities that depend on each other's existence: "To be entangled is not simply to be intertwined with another, as in the joining of separate entities, but to lack an independent, self-contained existence. Existence is not an individual affair. Individuals do not preexist their interactions; rather, individuals emerge through and as part of their entangled intra-relating" (Barad, 2007, p. ix).

For Haraway it is the experience of shared suffering that allows us to move beyond the mere representation of the suffering of the other. This experience enables learning, gaining new perspectives: in opening up "to shared pain and mortality" and learning "what that living and thinking teaches" (Haraway, 2008, p. 83). The ethical comportment of response-ability (ibid., p. 71) demands us to *respond* to what makes itself present (through the shared pain, through emergencies, through knowledge, through becoming-sensible towards another that enters our awareness).

Concluding Remarks

In the artistic examples I presented here, aesthetic experience is shared on the level of biochemical, atmospheric processes: the body's ongoing respiration produces carbon dioxide, whereby it becomes part of the living network in *H.O.R.T.U.S.*, or the symbiotic relationship in *BIO[lum]SKIN*. The reflection, representation and subsequent actions are extended, accompanied and partly enabled by animated material agency.

Together with animated material agents as technologies that expand our senses, that make us experience a world beyond our bodily or perceptual limits, human agency might evolve as something world-involving, that derives from the ways in which we establish interactions between ourselves and our biological as well as digital environments. As I have argued throughout this text, new perspectives on life, new ways of critical thinking and experiencing within global transformation processes are enabled, if non-sensual, non-conscious correlations are made tangible for a sensual subject. The artistic examples presented here do not emanate from a bifurcated picture of the human as being opposed to the world, a world that offers endless resources to the skilled hands of man or that in terms of crisis turns against it. They offer alternative subject positions in reconfigured environments that allow productive, life-affirming forms of life, able to transgress species boundaries and to account for the activities, the intra-actions of our bodies on non-cognitive levels. Expanding the sensual responsiveness towards living processes of others – of non-human beings, dust particles, light rays and so on – art can contribute to other forms of bodily knowledge that are needed in order to act response-ably in ever-changing environments.

Notes

1 www.iaacblog.com/programs/biolumskin.
2 The microbiome comprises all of the genetic material within a microbiota (the entire collection of microorganisms in a specific niche, such as the human gut). See http://www.nature.com/subjects/microbiome.
3 www.ecologicstudio.com/v2/project.php?idcat=3&idsubcat=49&idproj=115.

References

Barad, K. (2007). *Meeting the Universe Halfway*. Durham, NC, and London: Duke University Press.
Berleant, A. (1995). *Aesthetics of Environment*. Philadelphia: Temple University Press.
Bourriaud, N. (2002). *Relational Aesthetics*. Documents sur l'art. Dijon: Presses du réel.
Bradley E. C., & Holzman J. (Eds.). (2015). *Responsive Landscapes: Strategies for Responsive Technologies in Landscape Architecture*. London and New York: Routledge.
Carlson, A. (2002). *Aesthetics and the Environment: The Appreciation of Nature, Art and Architecture*. London & New York: Routledge.
Frost, S. (2016). *Biocultural Creatures: Toward a New Theory of the Human*. Durham, NC: Duke University Press.
Haraway, D. (2008). *When Species Meet*. Minneapolis: University of Minnesota Press.
Haraway, D. (2016). Tentacular Thinking: Anthropocene, Capitalocene, Chthulucene. *E-flux*. Retrieved from https://www.e-flux.com/journal/75/67125/tentacular-thinking-anthropocene-capitalocene-chthulucene.
Latour, B. (2017). *Facing Gaia: Eight Lectures on the New Climactic Regime*. Cambridge and Medford, MA: Polity Press.
Merleau-Ponty, M. (1968). *The Visible and The Invisible*, ed. C. Lefor. Evanston, IL: Northwestern University Press.
Morton, T. (2013). *Hyperobjects: Philosophy and Ecology after the End of the World*. Minneapolis: University of Minnesota Press.

10 Crafting Baroque Sound

How the Making of Organ Pipes Matters Artistically

Peter Peters

Introduction

A core theme in the debate on artistic research is how epistemic and aesthetic aspects of artistic making processes are interrelated. Artworks and performances that result from artistic research are not only relevant in terms of aesthetic experience but also as epistemic claims (Biggs & Karlsson, 2010; Borgdorff, 2010). Although these claims may have a discursive character, making art as research can also articulate non-propositional, sensory, embodied, and experiential forms of knowledge. In this chapter, I want to develop a reverse approach. My question is how these non-propositional forms of knowing that are at stake in artistic making processes can open up new dimensions, locations, and moments of the aesthetic. Or to put it in philosophical terms, how ways of knowing that are involved in artistic making processes create the conditions of possibility for normative judgements. In answering this question, I will follow an empirical strategy. Presenting ethnographic fieldwork on a recent pipe-organ building project, I will show how the making of pipes that sound as good as their historical examples involved learning from historical treatises, practising the casting of organ metal in historical ways, and voicing newly made organ pipes in such a way that their sound character approximates that of eighteenth-century pipes.

Of all musical instruments, the pipe organ has the longest history of innovation. Organ builders always incorporated new practices in music-making and composition, as well as new technologies. In the organ world, organs are therefore considered to be aesthetic and technological mirrors of their time (Snyder, 2002). This makes these instruments and the practices of knowing, making, and performing that revolve around them a strategic research site to explore interrelations of the epistemic, the technical, and the aesthetic. My case study is a project initiated by the Orgelpark in Amsterdam, the Netherlands. This privately funded concert venue opened its doors in 2007, and its mission is to give the pipe organ a new relevance in musical life. Located in a former church, it now has four large organs and a number of smaller instruments. The church's original instrument, a German Romantic Sauer organ built in 1922, was restored. A new French Romantic was built in 2009 and in 2012 a style copy of the oldest organ in the Netherlands was unveiled. This instrument is an attempt to reconstruct a Renaissance organ built by Peter Gerritsz in 1479 (Peters & Cressman, 2016).

To fill the perceived gap between the Renaissance organ and the two Romantic organs, the Utopa Foundation that funds the venue decided in 2013 to build a new organ. The aim was to build an instrument for music from the Baroque era, especially music by Johann Sebastian Bach. To meet this goal, recent insights and experiences from historically informed organ-building practice were taken into account (Davidsson, 2003). The decision was made that the sound of organs built by Zacharias Hildebrandt (1688–1757) should be a reference. Hildebrandt was a contemporary of Bach and worked in the same region as the composer (Dähnert, 1962). The actual building of the new Baroque organ, inaugurated in 2018, was done by a consortium of four companies.[1]

In this chapter, I follow Munetaka Yokota, a specialist in historically informed organ building, and the workers at Hermann Eule Orgelbau in Bautzen, Germany, in their attempts to craft the sound of Hildebrandt's pipes. Their work started by determining the design of the pipes of the various stops of the new organ. Not only were measurements of Hildebrandt pipes taken as a point of departure but also knowledge about ways of pipe-making in his time. Yokota's conviction that traditional pipe-making practices should be followed closely played an important role in how the pipes for the new organ were made. Central to Yokota's thinking is the idea that simply copying an aged object makes no sense. Instead, today's organ builders should try to understand how historical builders worked.

I focus here on the problem of judging the quality of organ pipes and their sounds. In doing so, I will look at the role of materials such as alloys, wood, and cloth, as well as the use of machines and tools in manipulating these materials. The anthropologist Tim Ingold has written extensively about the role of materiality and skill in making (Ingold, 2013).[2] He criticizes a view of the making of artefacts as imposing a mental image on the material world. In contrast, Ingold thinks of making as "a process of growth" in which the maker is placed in a "world of active materials" (Ingold, 2000, p. 3). I use some of these ideas to better understand how Yokota and the pipe-makers at Eule engaged with their materials and their tools.

Learning Along

In recent decades, the academic study of musical practices has become increasingly interdisciplinary. At the heart of the practice turn in musicology and music sociology that has been developed by scholars such as Taruskin (1995), Goehr (1992), Small (1998), Born (1995; 2013), Hennion (2015), Becker (2008), and Cook (2014) lies the idea that the meaning of music does not reside "in" the notes of a written score. Rather, as DeNora formulates it, it is to be found in "musical text and their appropriations, mediated by the ways that texts are framed through resort to familiar conventions, through what people say about them, through how they are used and through where they come to be sited" (DeNora, 2016, p. xiii). In studying music as a practice, scholars such as DeNora also highlight the role of materiality, as well as non-propositional and sensuous ways of knowing.

Recently, musical practices and their technologies have become a research subject in STS as well as in the related field of sound studies. In their Introduction to a special issue of *Social Studies of Science*, Pinch and Bijsterveld mapped the ways in which STS could inform the study of music and sound (Pinch & Bijsterveld, 2004). Thinking of musical instruments as technological artefacts, enables the study of how their meaning and use evolved in different practices or how certain aspects were kept stable while others changed (Bijsterveld & Peters, 2010; Bijsterveld & Schulp, 2004; Pinch & Bijsterveld, 2015; Pinch & Trocco, 2009). Investigating the building of the new Baroque organ in an STS register means following the collaborative work of the researchers and instrument builders involved in its making. What work did they have to do to convincingly claim that the sound of the new organ pipes shares the sound qualities of eighteenth-century pipes in Hildebrandt organs?

In June 2013, I became a member of the core team that would supervise the design and building of the new Baroque organ at the Orgelpark.[3] This offered me a position as researcher-in-residence to follow the building process from the very start to the organ's inauguration in 2018. I was present at meetings, made study trips with the other members to relevant destinations abroad, visited the various building sites such as the organ builders' workshops and the Orgelpark itself, and conducted interviews with the main actors in the project. My aim here is not that of the organologist, the historian of technology, or the musicologist. Rather, I see myself as an empirical philosopher in the domain of music. Following the actors is for me not only a way to understand the practices that they embody, but also of learning along with them, asking how they solve practical problems and make judgements. This allowed me to reflect on themes such as the epistemic value of musical practices and their relevance as forms of artistic research.

Local Knowledge and Materials

How did organ builders work in the past and of what relevance is this knowledge? Finding answers to these questions summarizes much of the work that Munetaka Yokota (b. 1952) did in the past forty five years. He became artist-in-residence at California State University in Chico in 1984. In the six years that followed, Yokota recruited part-time workers and volunteers from the university's student population and from the Chico community. All by himself, he taught them the knowledge and the skills to build a big organ in the eighteenth-century style of Silbermann and Hildebrandt. This approach, he explained, mirrored the practice that was common in the Middle Ages in Europe. Organ builders would come to a village or a town to work on site with the local craftspeople and use local materials. "The wood had climatized in the area and the workers already knew how to work with it" (Munetaka Yokota, interview, Orgelpark, Amsterdam, 16 May 2014).[4]

To explain the differences between the historical and contemporary organ building practices, Yokota draws a parallel with historical performance practice of Baroque music:

> This older music is a kind of outline. As a performer, you have to do lots of things that are not written on paper. Today's composers, depending on the style, write down their music in more detail. This also happens in construction work today, or in car making, or architecture. Architects make very precise drawings and pretty much everything is determined in design rooms. In order to do that, you have to have calculatable and predictable materials. But the old method was not like that. Design had to be some kind of rough outline. The reason for that was not that designers were lazy, but they knew that the actual material that they had to work with could be different. So they could not plan things into small details. That was left to the craftsman who actually worked with it. They could do a lot of judgement by just tapping on the material [*taps on the table with fingers*] and listening to it. In this way, they could choose whatever was best for the job based on qualities that cannot be written down in a design sheet. So the

> technical and artistic judgement was often made on site, in the workshop, and this required a lot of empirical knowledge from the craftsman. (Yokota, interview, 16 May 2014)

Understanding how historical organ builders worked became a guiding principle in the later work of Yokota. His research did not only rely on reading written sources but also on doing extensive experiments with reconstructing historical making processes. This process reconstruction does not aim at imitating but at learning. Imitating, Yokota argues, would be senseless precisely because one would only copy site-specific solutions that were developed under unique circumstances with local materials and tools. Understanding why historical builders developed certain solutions is thus key to making new pipes that sound as good as the old.

Casting the Pipe Metal

Making a metal organ pipe starts with casting sheets of metal on a casting bench. The metal is usually an alloy of tin and lead, although historical pipes also show traces of other elements such as antimony and copper.[5] After the sheet has been cast, its surfaces are scraped to make them flat. Front pipes are often polished as well.[6] In general, sheets are cast by pouring the melted lead–tin alloy into a wooden casting box which has a narrow opening at the bottom and sliding this box over a casting bench. Thus the melt will run out and spread into a thin layer covering the bench. The melt cools down and solidifies into a metal sheet. Important parameters in the historical casting process are the speed with which the metal cools down and the thickness of the sheet.

In November 2015, the project team visited the workshop of Eule in Bautzen, Germany, where the pipes for the new Baroque organ were to be made. Munetaka Yokota was present to share his experiential knowledge of historical casting methods with the craftsmen at Eule. In modern organ-building traditions, the casting bench is made of stone or wood and it is covered with a fire retardant cloth. In historical traditions, however, the casting bench was covered with linen cloth on stone or with a layer of sand or wooden board. During a project at the University of Gothenburg in the 1990s that aimed to build a copy of a northern German organ in the style of Arp Schnitger, Yokota and his team established that casting on sand gives the metal a completely different quality than when cast in modern ways (Speerstra, 2003). Due to the sand bed beneath the molten metal, the metal cools quickly, causing the pipe metal to become harder than modern pipe metal. These qualities matched surviving historical pipes (Yokota & Ruiter-Feenstra, 2003).

In the casting room, piles of unprocessed metal sheets in different alloys cover the floor. Next to the window is the casting bench. To the right of the bench is a small room with the melting pot and a pile of tin bars. Above the melting pot is a rail that holds a pan in which the melted alloy is transported from the pot to the bench. At the other end of the casting bench is a scraping machine. A rough metal sheet can be placed on a rotating cylinder. By moving a chisel along the metal surface while the cylinder spins, the surface of the sheet can be scraped to the desired thickness. The casting bench is covered by a fire retardant cloth with a layer of flannel beneath it. Yokota explains that it should not be used. Instead, he wants to cast on stone covered by "old style linen alone without any flannel beneath it" (Yokota, video of visit to Eule Orgelbau, Bautzen, 12 Nov. 2015). The stone makes sure that the metal cools down quickly, which gives it a hard surface. The linen is necessary to make room for small gas bubbles that would otherwise cause a bad surface structure. Casting will start the next morning at 7 a.m.

It is still dark when we walk from the hotel to the workshop of Eule Orgelbau. In the casting room, the melting pot has been heating the organ metal throughout the night. Two of Eule's employees are wearing their protective overalls and gloves. One of them is stirring. Inside I see the swirling surface of the molten metal. Its movements and appearance remind me of mercury. The new linen has been tightly strung over the stone beneath it. It is fixed with two wooden slats that also serve as rails to guide the casting box. The casting box that will be used this morning has two compartments. The metal will be poured in the first and when the casting begins, a valve will be opened so the metal can flow into the second compartment. From there it flows onto the bench when the casting box is slid from one end of the bench to the other. Attached to the lower end of the bench is a rectangular bin to collect the remaining metal.

One of the employees keeps on stirring in the pan, while the other measures the temperature of the alloy at short intervals with an electronic thermometer. The atmosphere in the room is one of concentration and expectation. Suddenly the two men tilt the pan towards the casting box and pour the metal in it (Figure 10.1). The sound of flowing fluid fills the room. The empty pan swings back. The two men quickly grab

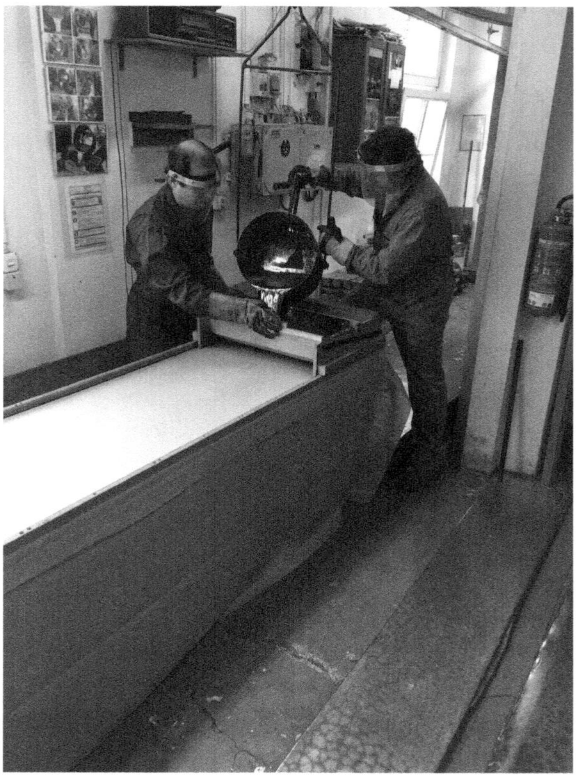

Figure 10.1 Pipe metal casting experiment at Eule Orgelbau, Bautzen. Photo by Peter Peters.

the casting box. Their bodies slightly bent, they walk a few steps along the casting bench, steadying the movement of the box that makes a whizzing sound over the linen. Behind the box, I see the shiny surface of the metal. At the end of the table, the remaining metal flows into the collecting bin, and spats of it land on the floor where they immediately solidify. The two men raise their protective head shields and look at the metal. Where it cools down its surface gradually becomes dull. Yokota and the two employees start discussing the quality of the metal sheet. They agree that the metal has cooled down too slowly because it was too hot. Yokota suggests to cast at a lower temperature. Several sheets are cast before the right temperature is established.

Yokota explains how eighteenth-century craftsmen were able to judge the right temperature of the molten metal. They knew the temperature relative to the procedure that they always followed, as well as the viscosity of the alloy.

> The alloy first has a completely liquid form, but when the temperature goes down, there is a kind of pasty, sticky form, and only then does it become solid. So when they are stirring the pot, they can pretty much judge the temperature. When it has this pasty form, the temperature stays there for some time. At this point of metamorphosis, they should stop waiting and just pour and go. (Yokota, video of visit to Eule Orgelbau)

Controlling the temperature of the cast metal is a complex affair, all the more so because the stone gets warm after several sheets have been cast. This has to be compensated for to get the right cooling coefficient, necessary for a good structure to the metal.

When the thickness of the newly cast metal sheets is measured, it turns out to be more or less equal. Yokota would rather see that the sheets were tapered: thicker at the start of the sheet and thinner at the end. From his previous experiments he knows that this makes it easier to follow the style of historical pipe makers. How to produce this tapered shape? With the two-chamber casting box this will not be possible, Yokota argues. He points at the scraping machine at the foot end of the casting bench:

> That kind of scraping machine was invented in the nineteenth century. It can produce a constant quality. The drum spins, and the operator moves the cutter sideways over the metal to make it thinner. A machine like this requires a metal plate of constant thickness from the beginning to the end. So the double box casting procedure is connected to that machine. (Yokota, video of visit to Eule Orgelbau)

To cast a tapered sheet, a single box is necessary. This type of box was used in eighteenth-century casting operations, as Yokota knows from historical organ-building treatises. When operated in the correct way, it will produce a tapered sheet. Because of the weight of the metal in the box, more will flow out at the start than at the end of the sliding movement when the box is nearly empty, resulting in a sheet that is thicker at the start and thinner at the end. The metal can thus be cast much closer to the final thickness. Yokota explains why this is important. In the eighteenth century, scraping had to be done by hand which was time and labour intensive. Hence it was rational to cast metal in a way that required less scraping: thin and tapered. With the coming of the scraping machine, sheets with equal thickness became the norm. This lead to problems, Yokota says. Thick metal casting means slower cooling down.

As a result of this the metal will become coarser in its grain structure which is undesirable. Casting thick means that the hardest and finest parts of metal will be scraped away, leaving metal of poorer quality (Yokota, video of visit to Eule Orgelbau).

Hammering Wind

Newly assembled organ pipes sound far from ideal.[7] Even if they have been built precisely according to specified measurements and scalings, the work of a voicer is needed to give them their final sound character, or voice. This work presumes a combination of knowledge, skill and manual dexterity, and a trained sense of hearing. To give a pipe its voice, the voicer manipulates all the parameters of a pipe that determine its sound character. In the case of a flue pipe, these are the size of the foot hole, the position of the upper and lower lip, the position of the languid (the metal plate separating the pipe foot from the pipe body, except for a small slot called the windway), the shape of the windway, the cut-up (height of the mouth in relation to the width of the mouth), and the structure of the languid edge or bevel. Making even very small changes to one or more of these parameters can result in a marked change in tone characteristics.

By determining the sound character of an organ, a voicer provides musicians with the sounds to make their music. This relation between sound, expression, and music lead to the idea, commonly shared in organ worlds, that voicing involves aesthetic judgements. As Segurado (2015) remarks, this judgement is more complex than distinguishing between a beautiful and an unpleasant tone. It relates to all the elements in the organ sound. Balancing and judging the various elements in the sound of a pipe are arduous and time consuming tasks that requires, often tacit, knowledge, skills, and judgements. The difficulty of documenting the process of voicing may be the reason that it is under-represented in historical treatises on organ building. This does not mean that organ voicing is simply a subjective affair. Precisely because voicing involves aesthetic judgements and reflects musical tastes, its history can be traced by studying surviving historical pipes and organs in relation to the aesthetic ideals of a period. Baroque organ builders favoured pipes that had a clear and pronounced speech and they compared their sounds to the human voice. In the Romantic era, many organ builders and organists preferred a cleaner, less noisy organ pipe tone. Understanding these historical conventions was an important aspect of the voicing work that Munetaka Yokota did on the new Baroque organ.

In his voicing of the pipes for the new organ, Yokota tried to respect Hildebrandt's example as much as possible, yet there were many different parameters he could change. "There is a certain amount of freedom. But if we deviate from that too much, then it is really out of Hildebrandt's range. I try to make a good theoretical guess, to recreate his language" (Yokota, recorded at Elbertse Orgelbouw, Soest, 29 Sept. 2016). Taking a mixture pipe as an example, Yokota shows me how he voices it. He blows on the pipe:

> Do you hear this, like scratching sound? I call this a "cough". This is created by a windway that is a little too wide [*blows again*]. And also wind is going outward too much. So in this case, I am going to close the windway just a bit [*uses a small hammer to give short, soft taps on the lower lip* (Plate 10)]. This reduces the cough [*blows again, and blows much harder to the point that the pipe overblows,*

> *i.e. sounds an octave higher*]. Maybe a bit more [*gives very small taps; looks at the pipe, and overblows it*]. Now I have to check the overbite.[8] And, eh, the languid position. I could make the languid position a bit lower [*takes a thin metal rod that he sticks into the pipe until it reaches the languid, puts the pipe against his body and hammers three times on the rod; overblows again*]. So when I blow very hard, overblow, that means that air went into the pipe a bit more. So this is just the right timing [*hammers on the corners of the lower lip*]. Now it is a little too early, so let's reduce the overbite to bring the air outward back a little bit [*hammers on the upper lip*]. I already checked this amount of overbite. This is important because that and the width of the windway and the height of the languid, those three elements, determine how the wind hits the upper lip. (Yokota, interview, Orgelpark, Amsterdam, 22 Dec. 2017)

There is still some "cough" left. Yokota observes that the speech is now very fast and it has a little accent at the beginning, "but it is of a good kind. So it can function as a mixture, and the sound is not too soft" (Yokota, interview, 22 Dec. 2017).

To voice the pipes, Yokota uses many tools: small knives, hammers, metal hooks and rods, and spatulas. He has made or adapted most of these tools himself "improvisationally", as he himself says. "I have to make tools first to do the job that I want to do [*laughs*]. And after thirty, forty years, I have a tool box full of homemade tools" (Yokota, interview, 22 Dec. 2017). He shows a small knife with a wooden handle and a triangular blade. The flatness of the blade is important, he explains, because it enables him to make very straight cuts in the pipe metal when adjusting the cut-up of the mouth. Originally, it was a shaving razor. One of his small hammers is normally used by clock makers. Yokota adapted the surface of its hammerhead. It is polished and shaped slightly round, and its edges are strongly rounded off. He has used this hammer for forty years and knows exactly where to hold the handle to give the kind of taps that he needs. Another instrument is made out of a file. The surface is flattened and its tip bent into a hook. It can be used to pull out the lower lip to widen the windway. Because it is very thin, it can be used for narrow windways.

When I observe Yokota during the pre-voicing of the pipes, it strikes me that judging the sound qualities of a pipe is only part of what he does. Examining the pipe's geometry and comparing it to the measurements that give the proportions of original pipes is just as important as an idea of how the pipe should sound. Some stops need a more "crisp" sound than others, he knows from his studies of historical sources but also from listening to historical organs. Does Yokota have a sound image of Hildebrandt's pipes in his head?

> Yes and no. I do have some image of Hildebrandt's organs in my head. On the other hand, the Hildebrandt organ sound we hear today may not be the original sound. I rely partially on what I hear, but partially on these geometrical relations. If we make the geometry right for the Hildebrandt sound and I use other knowledge, like what kind of treatment has to be done to the languid and so on, then the theoretically correct method is applied. Whatever the sound that comes out, I have to accept it as: this must be the original. … On the other hand, if it is obviously wrong or strange, first I doubt: my sense might be wrong. So it is a very complicated process to make this judgement. I hope to use my experience as a voicer or a researcher who has seen many old pipes, and some of these observations in

> the past inform this judgement. … Judging means including some acceptance of what sounds very strange, but what maybe should sound like that. Sometimes it is very painful and challenging. You see, because Hildebrandt is not myself. Hildebrandt might have liked the steak rare, I might like it more well done. But I have to accept his way [*laughs*]. (Yokota, recorded at Elbertse Orgelbouw).

In crafting the sound of Hildebrandt's pipes, Yokota starts from theoretical knowledge of original measures and scalings and tries to understand the geometry that results from these parameters. The resulting sound might not be what we like, but it is apparently what Hildebrandt himself wanted to hear. Making judgements and choices requires intellectual, tactile and sensory knowledge and skills. This kind of normativity, where respecting Hildebrandt's aesthetics might go against one's own intuitions and preferences, is what makes an artful skill.

How Machines and Materials Matter Artistically

In his book *The Perception of the Environment*, Tim Ingold puts the concept of skills centre stage as the action and perception of an organic being in a structured environment. This focus on continuous engagement challenges the opposition between intellectual design and mechanical execution, Ingold argues. Instead of being prior properties, intentionality and functionality are immanent in a practice itself. "Thus whereas in the artisan's handling of his tools, the movements of their working points are guided by his own perception, the motions of the machine, and any tools attached to it, are predetermined" (Ingold, 2000, p. 289). For Ingold, the history of human technicity can be understood as a 'history of externalisation' in which the artisan is moved from the centre to the periphery of production (ibid.).[9]

In *Making* Ingold revisits the theme of the relation between mental image and material object, now asking the question of what it means to make things. He contrasts two approaches. The first starts from the idea that making can be thought of as unifying materials provided by nature with conceptual representations springing from a cultural tradition. In this "hylomorphist" approach, Ingold argues, "practitioners impose forms internal to the mind upon a material world 'out there' " (Ingold, 2013, p. 21). The second "morphogenetic" approach views making as a process of growth. Here, the maker does not stand aloof from the world of active materials ready to receive the preconceived design but intervenes in worldly processes already going on (ibid.). We should think of making as *correspondence*, Ingold claims. To explain this concept, he gives the example of activities such as flying a kite, playing a cello, or using a potter's wheel. In the case of the latter, the maker is not imposing a mental form on matter, but the making emerges from the going together of gesturing hands, the rotations of the wheel, and the behaviour of the wet clay. The potter needs the wheel to correspond with the clay, just like the player needs the cello to correspond with the sound, or a flyer needs a kite to correspond with the wind. As Ingold puts it more abstractly: the kinaesthetic awareness of the maker (the potter) is converted by way of a transducer (the wheel) into a corresponding flow of material (clay) (ibid., p. 107). "To correspond with the world, in short, is not to describe it, or to represent it, but to *answer to it*" (ibid., p. 108). Ingold's ecological perspective on making helps us to better understand how Munetaka Yokota and the craftsmen at Eule worked. When Yokota proposed that Eule's employees change their working habits and their tools, he asked them to correspond with the metal in a different way. This required changing their tools, such as the casting bench and the casting box. Using these in an effective way required multiple casting attempts. Their speed of walking along the casting bench mattered since it determined the amount of alloy that would flow out of the box and hence the thickness of the sheet. The temperature of the alloy mattered because its cooling speed was related to the hardness of the metal. The single casting box mattered because its construction enabled the casting of a tapered sheet that would require less scraping, which in turn resulted in a sheet of better quality. These steps were not planned in advance; rather, every casting attempt might result in failure or success depending on the singular casting event. Similarly, when using his improvised tools during his voicing of the pipes, Yokota responded to their sounds and their materiality instead of imposing his mental image of it on them.

Conclusion

Casting organ-pipe metal and voicing organ pipes according to historical examples shows how non-propositional, embodied, and experiential forms of knowing that emerge in making processes open up new dimensions, locations, and moments of making aesthetic judgements. It is through these practical judgements that, in the end, the organ builders in Amsterdam were able to build a new Baroque organ the sound of which is considered to be artistically interesting and relevant to performing Baroque music. To function in a musically interesting way, the organ pipes had to be constructed according to specific measurements, their components had to be put together in a skilful way, and it has to be given a voice by combining skilled ears and hands.

I have shown how materiality, tools, and machines mattered in this making process. Based on my observations of the pipe making and voicing for the new Baroque organ, I argue that, as a process, it involved various normativities. In the case of the new Baroque organ these were not limited to subjective judgements, but included tracing, specifying, and constructing the relevant conditions under which normative judgements are thought to be relevant and valid. Yokota's approach to pipe-making is far more complex than simply re-enacting historical craftsmanship to discover lost meanings. He aims at understanding the "grammar" of Hildebrandt's pipe making, a certain logic that is at the same time local and situated, yet can be adapted to the new situation in the Orgelpark. His approach is hybrid in the sense that it combines insights and knowledge, instruments and tools from various contexts in order to create something that works in a new situation. Making an organ pipe sound good can only be done through creating a hybrid technical space of materials, machines, tools, and skills through which normative judgements and their reasons, be they intellectual, tactile, sensory, or aesthetic, can be articulated in new ways.

Acknowledgements

I want to thank Munetaka Yokota and Dirk Eule and his employees at Hermann Eule Orgelbau, Bautzen, for introducing me to their craftsmanship. I thank Henk Borgdorff, Hans Fidom, and Neil Smith for their comments on draft versions of this chapter. I am also grateful to the Utopa Foundation in Leiden for making my research on the new baroque organ project possible.

Notes

1 These companies are Elbertse Orgelbouw BV (the Netherlands), Munetaka Yokota (Japan), Hermann Eule Orgelbau (Germany), and Sinua (Germany).
2 Ingold's ecological approach is informed by James Gibson's view of perception as an achievement not of a Cartesian mind in a body but of an organism exploring its environment (Ingold, 2000, p. 3).
3 The board of the Utopa Foundation, that funds the Orgelpark, had decided that it would be beneficial to the project to have a social scientist on board with a background in STS and an interest in research into musical practices as well as artistic research. It was agreed that my research time would be funded partly by my university and partly by the Utopa Foundation.
4 It is ironic that local materials were used for the organ in Chico as well. To produce the alloy for the front pipes, lead from spent bullets from the LAPD shooting range was used (Yearsley, 2010).
5 The proportions of tin and lead can vary. Tin strengthens the alloy and is considered more beautiful than lead. Tin is also far more expensive than lead which is why the pipes standing in the façade usually have a slightly higher percentage of tin than pipes standing inside the organ case Because of their weight, pipes made of lead have to be carefully supported to prevent them from collapsing under their own weight (Shannon, 2009, p. 130).
6 Some organ pipes are made of lead only. The sheets for these pipes are usually hammered after they have been cast to make the metal structurally stable.
7 How a pipe sounds is determined not only by its physical characteristics but also by its spatial relation to other pipes and its position in the organ case. Another important determinant of the sound is the wind pressure that is set for the organ as a whole. Finally, the sound of a pipe depends on the room acoustics of the church or hall in which it sounds. A distinction is usually been made between pre-voicing and final voicing. Pre-voicing is done in the organ workshop. It entails checking if the pipe's components have been built and positioned according to right measures, cleaning and repairing imperfections that arose during the production, and making sure that the pipe behaves properly when producing its tone. The final voicing of the pipes is done when the new organ is placed in the room and all the pipes are placed in the organ case. Here the goal is to adjust the sound of the pipes to the other pipes in a rank and in the organ and to adjust the sound to the acoustical characteristics of the room.
8 "Overbite" is a characteristic of Hildebrandt pipes. The upper lip is not aligned with the lower lip, but slightly pronounced.
9 This argument resonates with Sennett's analysis of the changing role of crafts in modern societies (Sennett, 2008).

References

Becker, H. S. (2008). *Art Worlds*. Berkeley: University of California Press.

Biggs, M., & Karlsson, H. (Eds.) (2010). *The Routledge Companion to Research in the Arts*. London: Routledge.

Bijsterveld, K., & Peters, P. F. (2010). Composing Claims on Musical Instrument Development: A Science and Technology Studies' Contribution. *Interdisciplinary Science Reviews, 35*(2), 106–121.

Bijsterveld, K., & Schulp, M. (2004). Breaking into a World of Perfection Innovation in Today's Classical Musical Instruments. *Social Studies of Science, 34*(5), 649–674.

Borgdorff, H. (2010). The Production of Knowledge in Artistic Research. In M. Biggs & H. Karlsson (Eds.), *The Routledge Companion to Research in the Arts* (pp. 44–63). London: Routledge.

Born, G. (1995). *Rationalizing Culture: IRCAM, Boulez, and the Institutionalization of the Musical Avant-Garde*. Berkeley: University of California Press.

Born, G. (Ed.) (2013). *Music, Sound and Space: Transformations of Public and Private Experience*. Cambridge: Cambridge University Press.

Cook, N. (2014). *Beyond the Score: Music as Performance*. Oxford: Oxford University Press.

Dähnert, U. (1962). *Der Orgel- und Instrumentenbauer Zacharias Hildebrandt; sein Verhaltnis zu Gottfried Silbermann und Johann Sebastian Bach*. Leipzig: VEB Breitkopf & Härtel Musikverlag.

Davidsson, H. (2003). Organ Building in Northern Europe since 1969: Historical Revival and Renewal. In J. Speerstra (Ed.), *The North German Organ Research Project at Göteborg University* (pp. 329–340). Gothenburg: Gothenburg Organ Art Center and Gothenburg University.

DeNora, T. (2016). *Music-in-Action: Selected Essays in Sonic Ecology*. London: Routledge.

Goehr, L. (1992). *The Imaginary Museum of Musical Works: An Essay in the Philosophy of Music*. Oxford: Oxford University Press.

Hennion, A. (2015). *The Passion for Music: A Sociology of Mediation*, trans. M. Rigaud & P. Collier. Farnham: Ashgate.

Ingold, T. (2000). *The Perception of the Environment: Essays on Livelihood, Dwelling and Skill*. London and New York: Routledge.

Ingold, T. (2013). *Making: Anthropology, Archaeology, Art and Architecture*. London and New York: Routledge.

Peters, P. F. & Cressman, D. M. (2016). A Sounding Monument: How a New Organ Became Old. *Sound Studies: An Interdisciplinary Journal, 2*(1), 21–35.

Pinch, T., & Bijsterveld, K. (2004). Sound Studies: New Technologies and Music. *Social Studies of Science, 34*(5), 635–648.

Pinch, T., & Bijsterveld, K. (2015). Instruments and Innovation. In J. Sheperd & K. Devine (Eds.), *The Routledge Reader on the Sociology of Music* (pp. 301–308). London and New York: Routledge.

Pinch, T., & Trocco, F. (2009). *Analog Days: The Invention and Impact of the Moog Synthesizer*. Cambridge, MA: Harvard University Press.

Segurado, J. (2015). Never Heard Before: A Musical Exploration of Organ Voicing (unpublished PhD thesis). University of Gothenburg.

Sennett, R. (2008). *The Craftsman*. New Haven, CT, and London: Yale University Press.

Shannon, J. R. (2009). *Understanding the Pipe Organ: A Guide for Students, Teachers and Lovers of the Instrument*. Jefferson, NC: McFarland.

Small, C. (1998). *Musicking: The Meanings of Performing and Listening*. Middletown, CT: Wesleyan University Press.

Snyder, K. J. (Ed.) (2002). *The Organ as a Mirror of Its Time: North European Reflections, 1610–2000*. Oxford and New York: Oxford University Press.

Speerstra, J. (Ed.) (2003). *The North German Organ Research Project at Göteborg University*. Gothenburg: Gothenburg Organ Art Center and Gothenburg University.

Taruskin, R. (1995). *Text and Act: Essays on Music and Performance*. Oxford: Oxford University Press.

Yearsley, D. (2010). The Organ-Building of Munetaka Yokota. *CounterPunch*, 3 Dec. Retrieved from https://www.counterpunch.org/2010/12/03/the-organ-building-of-munetaka-yokota.

Yokota, M., & Ruiter-Feenstra, P. (2003). Historical Pipe Metal Casting Techniques in Seventeenth Century North Germany. In J. Speerstra (Ed.), *The North German Organ Research Project at Göteborg University* (pp. 165–186). Gothenburg: Gothenburg Organ Art Center and Gothenburg University.

Part III

Experiments

11 Everything Will Be Screen

Readdressing Screenness through Art-based Experiments

Claude Draude

Noticing Screens: Introducing Screenness

This chapter addresses the pervasiveness of computer screens by following a phenomenological conception of screens as mattering only as screens-in-the-world. In this first section, the term "screenness" is introduced to describe how artifact, enactment and contexts of actions need to be dealt with in a holistic manner. Following this, the specific characteristics of the computer screen, interweaving basic principles of computing and cultural impact, are discussed as grounds for further consideration. This discussion and the phenomenological conception of screens provide the basis for experimenting with screenness through art-based research. The chapter closes by summing up findings from the experiments. Furthermore, the question of the shaping of knowledge through the suggested methodological reframing is discussed.

As a means for providing information, for entertaining or for aesthetic purposes, screens have mattered in Western culture at least since the Renaissance. As varied as screens are in terms of displayed content, materiality or cultural-historical setting, they share common characteristics that make them recognizable as screens. Lev Manovich (1995) points out that from paintings to the cinema, a screen is a "flat, rectangular surface ... intended for frontal viewing ... it exists in our normal space, the space of our body, and acts as a window into another space." Manovich adds that the proportions of screens have not changed throughout history either and that it is typical for the representational space that the screen provides to be of different scaling than the person's "body space".

Over the last decades, computing technology has changed the functionality, as well as the occurrence of screens significantly. I write this chapter using a notebook computer. As I type, letters appear immediately on the screen; incoming e-mails pop up at the margin of my field of vision; I alternate between different "windows" as I use text editor, web browser, bibliography software and various communication tools. The computer system notifies me of changes in system state and announces alerts. The display constantly demands and holds my attention. The human–computer interaction set-up of monitor and keyboard is a longstanding one for knowledge and office work.[1] With the development of computers and advances in mobile technology, the ubiquity of displays[2] has reached a new level. We now live in a pervasive screen culture. Smart phones and handheld devices are changing the way people communicate and connect with each other and their environment. Video-call software provides real-time windows into somebody else's world. The real-time and interactive character sets computer screens apart from paintings, photography or cinema. The latter supply a window into past events and not into the immediate present.

Furthermore, augmented reality applications, like the Pokémon Go mobile game (Niantic, 2016) or museum exhibition software like Skins and Bones (Smithsonian National Museum of Natural History, 2015), let the human users experience a so-called mixed reality of computer-generated audiovisual overlays that merge with the physical world through the screen. Other examples are wearables and smart watches that collect biometric data. Their small displays give computerized insights into the human body and report constantly on its performance. Wearables can connect gathered data to others: for example, to fitness communities or sports attire companies. This invokes another meaning of the word "screen" as in "to screen" or "to monitor".

Notable for computer displays is that the screen not just serves as a medium for output but as one for input as well. A famous historical example of this is the USA Cold War military system SAGE, which stands for Semi-Automatic Ground Environment (Enticknap & Schuster, 1959). SAGE consisted of large computers and networks that produced a unified image from radar-site data. Operators used light guns as input devices to select targets on screen and initiate commands to attack.[3]

With today's increase in mobile, smaller and lighter technological devices, the screen as input instrument has become of widespread use. Smartphones do not have separate keyboards – we type and interact directly onto or through the screen.[4] In our experience, the screen becomes the computer. Touch screens are also getting more common for laptops, desktop workstations and shared work environments like multi-touch tables. Noteworthy is how capacitive touch screens work by altering the electrical field once the finger touches the screen. The finger partakes in the electrical circuit because the human body conducts electricity (Barrett & Omote, 2010). The human is part of the screen interface.

Despite the ubiquity of screens and the advances in technology, the lack of attention the screen as such receives in research, as well as the scope of most screen studies, have been criticized (Ziewitz, 2011; Huhtamo, 2004; 2009; Kuhn, 2009). There are, of course, exemptions to the stated lack, like the body of work on urban screens exploring media architectures, communal space and participatory practices (Krajina, 2009; McQuire, Martin & Niederer, 2009). For the computer screen, however, Malte Ziewitz (2011) states that screen studies address the topic in a dualistic manner, re-invoking the split between human and machine rather than analyzing the setting.[5] Instead, he suggests drawing on STS's long tradition of tracing how subject–object relations emerge as network[6] or rhizomatic activities (Beetz, 2016). In particular, Ziewitz takes up Annemarie Mol's influential ethnographic study on health services in a Dutch town and her findings demonstrate how objects are being enacted in and through practices of varying contexts (Mol, 2002, p. 5; cited in Ziewitz, 2011, pp. 209–210). The epistemological shift that has been articulated as material(-)semiotic by (feminist) STS and technoscience scholars allows us to understand how objects stabilize in inter- or intra-action rather than presume they occur as predefined entities. Donna Haraway finds that "bodies as objects of knowledge are materialsemiotic generative nodes ... 'Objects' like bodies do not pre-exist as such" (Haraway, 1992, p. 298). John Law has named actor–network theory "a disparate family of material-semiotic tools", which "describes the enactment of materially and discursively heterogenous relations that produce and reshuffle all kinds of actors"

(Law, 2007, p. 2). Screen studies, against the background of a material-semiotic reframing, then calls for situated, enacted, screens-in-practice research. Following this, Ziewitz finds, that

> a focus on enactment does not go well with a focus on objects – even if conceptualized as multiple. The challenge of attending to screens as "objects of interest" may thus be better understood as an invitation to engage in and interfere with ongoing enactments through the concept of "screens". Otherwise, one runs the risk of seeing screens everywhere and nowhere at all. (Ziewitz, 2011, p. 223)

I follow Ziewitz' argument of screens-in-practice up to the last part of the quote. His remark of "seeing screens everywhere and nowhere at all", I do not read as mutually exclusive as it implies. Instead, I consider it as describing a field of tension between "everywhere and nowhere" which is constitutive for current screen culture whose enactments oscillate between modes of in/visibility, non/perceiveability, absence/presence/telepresence and which are only insufficiently described by either/or options.

To explore this tension further and to address screens in a non-dualistic way, the concept of screenness or screenhood elaborated by Lucas D. Introna and Fernando M. Ilharco (2006) is helpful. Their approach is based on Martin Heidegger's[7] phenomenological and strictly non-Cartesian understanding of *Dasein* (being-in-the-world):[8]

> Apparently the screen enters our ongoing involvement in-the-world – as a screen – when we attend to it by turning it on. When we push the "on" button the screen captures our attention as it is the place, the location, the setting, the scene, in which what is supposedly relevant for us at that particular time is happening. Screen has as its necessary condition this supposed relevance. ... Yet, this capturing and shaping of our attention, that screens are, does not sometimes happen (and sometimes not), i.e., it is not only when we push the "on" button that screening is present. On the contrary, *that* we push the "on" button means precisely that the screening of screen – its possibilities as well as its transparency – is already there as a horizon of possibilities. As beings already in a "screening world" we are already relying on, and basing ourselves and our possibilities for being on, this very *screenhood* of screens. (Introna & Ilharco, 2006, p. 63)

Thus, screenness is already inextricably enmeshed in our enacting of and interacting with screens. It "conditions our engagement with certain surfaces in as much as we comport ourselves towards them as screens" (Introna & Ilharco, 2006, p. 64). When undertaking screen studies this means that being-in-the-world is tied to the experience of screens-in-the-world. Thus, the experience of encountering screens as objects or technological entities is stabilized, or becomes intelligible, only through their enactment in contexts in which they already make sense.

What Matters Through the Screen

Above, I have pointed at special characteristics of the computer screen. In contrast to the more classic screens of Renaissance paintings, photography or even the dynamic screen of the cinema, computer displays serve as interactive, input/output devices. Furthermore, they are windows into (tele)present worlds, and they provide windows into fictive or even impossible worlds. Certainly, the latter is true for cinema, paintings and photography, too. The promise of computing, however, is that these worlds can be entered or at least be experienced in an interactive way not realized by traditional media. This distinctiveness of the computer screen is most obvious in augmented reality or virtual reality scenarios. Here, the overcoming of the screen as barrier – or the dissolving of the screen – is a major narrative thread. Interestingly, this seems to fit a non-dualistic perspective on screen culture, because it favors connectedness of human user, context and technological device over their separation. In the following, a short overview of the computer screen as facilitator or gateway of/to "other worlds" is given.

In 1965, Ivan E. Sutherland, forerunner of graphical user interfaces and artificial-intelligence systems, wrote that "a display connected to a digital computer gives us a chance to gain familiarity with concepts not realizable in the physical world. It is a looking glass into a mathematical wonderland." He continues by describing his vision of the "ultimate display", which "would, of course, be a room within which the computer can control the existence of matter. A chair displayed in such a room would be good enough to sit in. Handcuffs displayed in such a room would be confining, and a bullet displayed in such a room would be fatal. With appropriate programing such a display could literally be the Wonderland into which Alice walked" (Sutherland, 1965, p. 508).

With this, Sutherland invokes the story of *Through the Looking-Glass, and What Alice Found There* (Carroll, 1992). Like the *Chronicles of Narnia* (Lewis, 2000) and other classic tales, Alice's adventures are stories of passage and transition, of overcoming a threshold where the protagonist leaves the known world behind and arrives in a fantasy world. In this world, familiar objects, actions and identities get rewritten; new ones get invented. In *Through the Looking-Glass*, Alice manages to slip through the living room mirror into the "mathematical wonderland" of a chess board.

In Carroll's story as well as in Sutherland's vision, an optical, technological device (the mirror/display) serves as a gateway to a new world (fiction/virtual, augmented or mixed reality). The unique quality the gateway possesses is the power to overcome the constraints of the given physical reality. In both scenarios, computer screen and story line, passing the threshold is tied to semiotics. For the written version of Alice,[9] the mirror dissolves through the power of text, whereas Sutherland's ultimate interface promises that computer code will produce and control new forms of materiality. For this chapter, I briefly highlight two of the many intra-acting[10] threads drawing through the fabric of these material-semiotic transformations: "doubling/reflection" and "simulation/diffraction", which define modes of representation, embodiment and spatiality on/off/beyond the screen.

Doubling/Reflection

At first glance, the living-room mirror in *Through the Looking-Glass* simply doubles the living room and Alice through its reflective quality. Alice, however, questions the quality of the objects represented on the screen and the mirror-world it produces. She asks, whether the fire in the fireplace on the other side can keep her warm or if it is only "pretence" (Carroll, 1992, p. 112). Eventually, the materiality of the mirror changes, it softens, Alice can get through and explores the new world.[11]

Simulation/Diffraction

Following this, the story of Alice starts with the paradigm of doubling and reflection but leaves it behind after the entrance scene. The world through the looking-glass diffracts rather than reflects Alice's reality: "Diffraction does not produce 'the same' displaced, as reflection and refraction do. Diffraction is a mapping of interference, not of replication, reflection, or reproduction. A diffraction pattern does not map where differences appear, but rather maps where the effects of differences appear" (Haraway, 1992, p. 300).[12]

Sutherland's vision of the ultimate display places the power to create a walk-in wonderland, produced by and beyond the screen and accessible through it, in the realm of computing. The computer's character as a special machine lies in its power to process code and to generate artifacts that appear animated. Alice's questions about the quality of Wonderland's objects and beings, as well as the identity and embodiment crisis that happens to her in the story, are very similar to issues that have been of concern in computing and in artificial intelligence or virtual reality, in particular. Usually, modeling in computing aims at mirroring or representing parts of the world (topics, objects, procedures etc.). But like Alice's looking-glass, the computer is no simple mirror either.

Interface design exemplifies the practices of computer science where the programmer invariably must handle processes of abstraction and de- and re-materialization. For anything to become an object of computing this means undergoing phases of semioticizing, formalizing and algorithmizing (Nake, 1993; Nadin, 2007). The human–computer interface, in a most basic sense, is a place for sign/symbol/signal transformation – a co-constructive production site. Considering the basic principles of computing, a field of tension between materiality and semiosis is produced. Oversimplified this means that, on the one side, there is the computational working mode of signal processing and, on the other side, there is the human user who understands certain symbols, signs and natural language.

Simulation, and trust in it and its believability, is tied to these transformative processes. Both, Alice's Wonderland and Sutherland's ultimate display, however, exceed the term simulation. Sybille Krämer describes the computer as *Virtualisierungsmaschine* (machine of virtualization) – and, using the mirror metaphor, as "ein interaktiver Spiegel dynamisierter Symbolwelten" (an interactive mirror of dynamized symbolic worlds) (Krämer, 2011, p. 313). In addition, Elena Esposito elaborates on the relation between simulation and virtuality. Simulation, she states, has received the role of substitution, of pretense, of "as-if". It still references the physical world. In contrast, virtuality refers to an alternate world that contains objects that count as real in this world. As a concept, virtuality cuts the link to other realities (Esposito, 2003, p. 270). What we find in today's screen culture, certainly, is no cut to other realities. Introna and Ilharco stress that "real life is not just another window, it is the *only* window, unless of course we treat the life on the screen as a game" (Introna & Ilharco, 2000, p. 71). Instead, what Manovich calls transcoding, a term which in computing means converting files from one format of encoding to another, occurs. Manovich uses transcoding to describe the co-evolutionary character of the "computer layer" and the "cultural layer" (2001, pp. 47, 63).

But what does this mean for a reconsideration of screens and screenness? Introna and Ilharco take up the mirror metaphor as well: "We might suggest that screens, as focal surfaces that grab and hold our attention, may indeed also appear to us as 'mirrors' of truth – however, not reflecting that which is before them but reflecting a way of living already implied in their screening" (Introna & Ilharco, 2000, p. 68). Following this, screens as human–computer interfaces, which a large number of current screens are, demonstrate the pervasiveness of basic principles of computing in today's culture. Means of information and communication, forms of knowledge and representations or constructions of embodiment and identities increasingly fall under the logic of what Haraway notedly calls "the translation of the world into a problem of coding" (Haraway, 1992, p. 164).

Methodological Reframing

In the following, I provide insights into a methodological reframing or, put differently, a continuation by other means of researching screens and addressing screenness. The suggested approach aims at promoting and performing non-standard ways of knowledge production in academia or in a technological field. Furthermore, it aims at contributing to the development of new methods for rethinking and reconstructing technological artifacts. The art-based research approach is grounded in phenomenology and takes up findings from New Materialism. As a basic epistemological question, the shaping of knowledge through research settings – available means, media, procedures, contexts, materialities – is considered.

In a publication on art-based research, Cathryn Vasseleu elaborates on the special quality of "translucency" regarding character animation in computing, where translucency must be formulated algorithmically. Vasseleu starts by giving a definition of translucency, as

> an incomplete transparency, or partial opacity. It is a liminal quality, existing on the threshold between clarity and obscurity. Optically, a translucent substance is one that is capable of transmitting light, but also causes sufficient diffusion of that light to prevent the perception of clear and distinct images through it. (Vasseleu, 2013, p. 158)

Vasseleu expands her analysis using works from phenomenological philosophy, Maurice Merleau-Ponty and Henri Bergson, as well as discussing physical and digital works regarding translucency. Referring to Merleau-Ponty's anti-optical approach of visual perception (Vasseleu, 2013, p. 159), Vasseleu stresses the agency and animating quality of how translucency intra-acts with matter, reconceptualizing her earlier definition of translucency derived from optics. A phenomenological approach regards visual experiences as being-in-the-world, as connectedness rather than as a human subject looking at an object. In this context, translucency is constituted by materiality, and perceiving it is an embodied, situated, referential experience. We do not see through a stained-glass window despite its stains, but because of its stained quality our conception is altered.

In accordance with her phenomenological elaborations, Vasseleu discusses a computational and a physical model of translucency. For the computational model, Vasseleu refers to Craig Donner and Henrik W. Jensen (2006), who test the validity of their algorithm by comparing photographic images of milk and of human skin tones to computer-generated simulations. The aim of the computational model is to motivate the same "visual response" as the photography of the physical object produces (Vasseleu, 2013, p. 161). Hence, embodiment and referentiality are especially important. The simulation of light entangles digital media object and physical human existence.[13] Vasseleu states that

> when viewers experience an image of light interacting with matter they attribute the reality of a carnal presence to it. Any difference between the simulation and physical enactment of the visual effect is, intentionally, undetectable at a conscious level. The image behaves like a voluminous object with communicable physical properties. How might we begin to pose what is unseen and unthinkable in that? (Vasseleu, 2013, p. 156)

To pursue this question further, Vasseleu, discusses Rachel Whiteread's work *Water Tower*[14] as a physical model. For this, the artist rendered the space inside one of New York's roof-top water towers as a "translucent, hollow, sculptural object cast from resin". The original tower that had served as the mold, was destroyed in the process. The sculpture by Whiteread was installed in the place of the original tower, providing a "ghostly imprint of the inside surface of the original water tank's interior" (ibid., p. 165). Remarkably, the translucency that the art work produces was not an observable quality of the solid former water tower. Whiteread's installation alters the visibleness of a common urban site. Thus, *Water Tower* serves as a means for discovery of what is commonly overlooked in familiar structures (ibid., p. 168).

Vasseleu uses both examples, the computational model with its simulation of milk and human skin and the resin cast sculpture of the water tower, to trace concepts of modeling and simulation. Considering the computer model as well as the physical sculpture allows her to open a space for "an unthought knowledge of physicality that invisibly shapes perceptual experience" (Vasseleu, 2013, p. 169). Vasseleu's quest for finding a way to address what is left unseen and unthinkable in the way computing technology reconfigures knowledge and artifacts motivates my own experimental research. Transferred to screens, this means asking what is left out in the current culture of screenness and tracing topics or modalities that are unattended. Considering the preliminary remarks on phenomenology, the question is to what kind of surfaces do we comport ourselves as screens. As I have stated above, Ziewitz' warning of seeing screens everywhere and nowhere at all[15] serves as a productive field of tension guiding my art-based experiments.

In the previous section, I have identified "doubling/reflection" and "simulation/diffraction" as nodal points in the role of the screen against its background of basic principles of computing. The dialectic character of the screen of "being everywhere/being nowhere" points to another important trait, the "absence" and the "presence" of screens or their (un)noticeability. Screens appear to be "silent objects". They reside in the background until they come to life. Mobile displays demand attention through push-up notifications, they vibrate and light up. Projection screens catch the eye, once the projector is lit and an image appears. It is usually not the screen as such we notice but the content it provides. The guiding principle for my art-based research is experimenting with the noticeability and visibleness of screens as screens.

I named this section "Methodological Reframing" and added that I want to pursue a continuation of researching screens by other means. By doing so, I run the risk of posing art as an opposition or as the Other, of science – which is not my intention.

What interests me is what happens when I attend to screens through different material practices than those commonly used in computer science (text-based research, writing, scientific experiments, software/hardware-construction).

The relation between art and science is prominently discussed in numerous publications. Useful for my approach are findings that stress the connectedness of art and science rather than their separation. Henk Borgdorff (2006) distinguishes between research "on the arts", "for the arts" and "in the arts". The latter does not assume an epistemological separation between art and theory. Kathrin Busch, referring to philosophy and art paradigmatically, states that:

> Art and theory, in effect, are nothing more than two different forms of practice interrelated through a system of interaction and transferences. In this constellation, philosophy neither brings the arts to the point nor does art sensualize philosophical truths; philosophy serves a knowledge-based artistic practice as a point of reference, similar, conversely, to how art might affect theoretical practice. (Busch, 2009, p. 1)

Julian Klein (2011) traces the art and theory dichotomy back to modernity and its formation of distinct scientific disciplines (cf. Latour, 1993). In contrast, he stresses that research as a quest for knowledge is a common denominator for both art and theory (or science). Klein characterizes art as experience – as a sensual, emotional, embodied process of gaining knowledge.

This resonates with older (Dewey, 1934) and contemporary perspectives. Barbara Bolt (2006), for example, uses the term "material productivity" to strengthen the agency of material settings and the process-like character of art as research.

Accordingly, the focus of my work is not the artistic product or object as such but thinking, reflecting and perceiving through art-based experimental set-ups, with a special interest in embodiment and site-specific situatedness. The notion of "experimenting", a term closer in origin to the natural sciences than to the arts, is key for this method of knowledge production through artistic research practice. Plausibly, Borgdorff (2013), in his rethinking of the epistemic status of artistic experiments, turns to Hans J. Rheinberger's elaborations of experimental systems in the life sciences (1992). Despite differences between artistic and scientific experiments – for example, in terms of reproducibility, generalization and controllability (cf. Borgdorff 2013, pp. 115–116) – Rheinberger's notion of experimental systems as *Orte der Emergenz* ("places of emergence") (2012, p. 7) holds true for both fields, as he himself finds. Rheinberger states that experiments function as *Überraschungsgeneratoren* ("generators of surprise") and *Maschinen zur Herstellung von Zukunft* ("machines to fabricate the future") in the arts as well as in the sciences (ibid., pp. 7–8).[16] Most useful is his definition of experiments as providing a systemic structure that, on the threshold between knowing and not-knowing, forms as space where new findings become possible (ibid., p. 8).

My art-based research draws on this concept of experimentation as a search movement. Michael Schwab states that: "science appears in artistic research like any other thing, not yet understood and probing its strength" (2008, p. 2). Thus, within my experimental set-ups, artistic practice serves as a generator and transformer of knowledge.[17] Accordingly, the set-ups serve as means of recontextualizing and defamiliarizing screens. In the following section, I present three of my art-based research experiments. The two older ones are located within more traditional art contexts, one in a curated show, the other in a gallery, whereas the newer work is part of my on-going field-based research. A discussion of the experiments follows in the subsequent closing section.

Experimenting / Experiencing Screenness

"I Think I Have Lost You"

Figure 11.1 shows one section of an installation in a former warehouse, which had been out of use for some time before it was turned into a temporary art space. As a part of the site-specific work, I turned the double-glazed windows into display cases. The bottom half of the outer window parts are glazed using a milky substance and black tea, which produces translucency. Plant parts, that have been collected from the adjoining park, are arranged inside the hollow area between the two glass panels. The upper parts of the windows are filled with sand from the construction site in front of the building. During the exhibition time, the sand gradually shifts, altering the amount of light that gets through, as well as the pattern it produces. There is no electric lighting inside or outside the room, therefore the shapes that the display cases

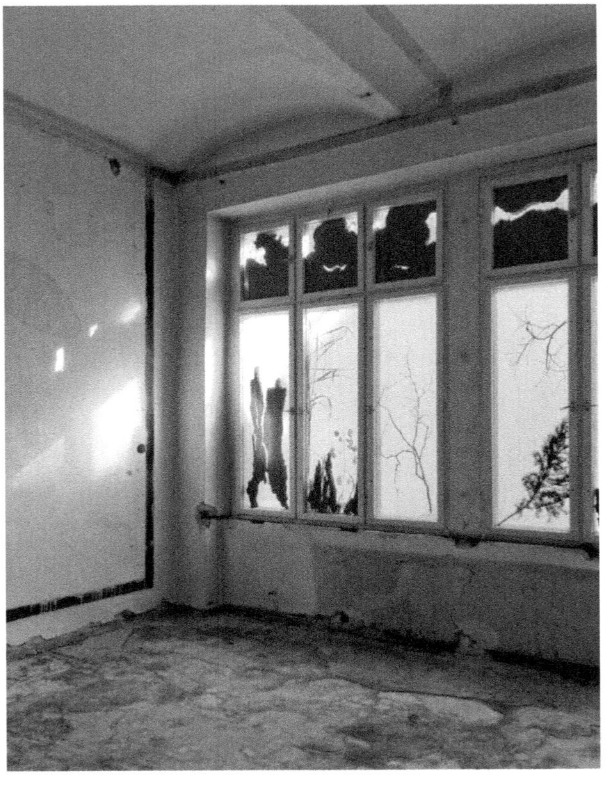

Figure 11.1 Claude Draude, I Think I Have Lost You (July–August 2010). Mixed-media installation in multiple parts and performance. Exhibition Ganze / Teile, Schuckerthöfe, Berlin.

cast, differ depending on weather conditions and time of day. When the sun shines, the display projects multicolored shapes onto the walls and floor. On cloudy days, the colors of the display itself seem to disappear. Starting at sunset, the patterns begin to dissolve, resulting in their disappearance when it is a dark, moonless night. From outside of the building, the plant parts appear blurred; details are hardly visible due to the glazing. Interestingly, the windows provide a three-dimensional display, but from the outside it looks like a flat screen. From inside the room, the plant parts invite a closer observance of the display. They motivate touching, but the glass serves as a barrier for touch – only light gets through. When the sun illuminates them, images are projected onto walls and floor that subsequently become screens as well. Thus, the windows serve both as a barrier and as link between the inside and the outside of the building, as well as they provide a space themselves and function as projection device. Turning them into display cases alters their visibleness: from something we ordinarily look through, they become something which is looked at.

"Everything is Taken"

Plate 11 shows a still from a film of the installation "Everything is Taken". A plant is exhibited away from daylight in the usually unused basement of the gallery. A slide projector illuminates the plant and part of the wall. It is not immediately perceptible that the projector shows an image of the plant that has been taken a few days earlier. The plant is turned into a screen for its own image. At the start of the exhibition, the plant and the picture of it coincide. But even at this stage, object and image diverge depending on the perspective of the viewer. The plant as a material object gets in the way of its previously taken image, while the light projection produces a shadow image of the plant as an additional overlay. During the time of the exhibition the divergence of recorded image and the appearance of the plant (due to lack of daylight) proceeds. The physical plant no longer fulfills its place as a screen. Instead, the wall behind the plant serves as a projection screen for the slide picture. The physical plant is still part of the experiment. As its appearance changes over time, new shapes are produced through the overlay of projector light and plant. In effect, a three-dimensional screen consisting of perishing plant, shadow images of the plant and the slide picture of the original state of the plant emerges.

"Everywhere, Nowhere"

Plates 12a and 12b are film stills from recordings taken in a public park. On warm summer mornings, an automated sprinkler system waters the park's lawn. The sprinklers are spatially distributed, and the sprinkler's tops move in coordination to cover a maximum amount of greenery. Depending on the alignment of the water jets and the spatial position of the viewer various visual effects occur. The first picture demonstrates different states the water produces from larger drops to a fog-like curtain. At a specific point in time, the crossing of the two sprinklers results in overlays and diffractive patterns. At their peak activity, a space emerges that obscures a clear vision of the opposite side of the lawn. People seen through this space look different. They appear as shadows or even ghost-like. The prism effect of the water jets makes physical properties of light visible. The translucent screen that is produced is constantly moving and fleeting. The second picture, taken from a different angle, makes the screening quality of the water jets more obvious. At a certain spatial position and with adequate sunlight, a tree casts a shadow on the water surface. In effect, the shadow renders the water transparent (in contrast to the translucency of the adjacent water jets). The small child in the background which was formerly obscured by the water screen is now visible. Besides the visual, the water sprinklers provide audio and haptic impact as well. On hot days, children play with the sprinklers; they run through the water screen and use its cooling effect. But even when you do not step inside the sprinkler space, the way the water changes the air and the sound it makes is noticeable.

Conclusion and Further Questions

I started this chapter with a discussion on the ubiquity of computer screens in the context of a phenomenological conception of screenness. Subsequently, I outlined basic characteristics of computing technology that characterize the cultural impact of the computer screen. As an interface or even as an entry point into a mathematical wonderland, the computer screen oscillates between modes of reflection, diffraction and simulation. For the methodological reframing, I followed Vasseleu's lead of discussing a physical model to evaluate her research topic of translucency. Thus, for my art-based experiments I purposely took screens out of the logic of computing. Questions of initial interest were: What is unseen or unattended to in computer screen culture? What kind of surfaces are enacted as screens and in what way?

In their set-up, the art-based experiments form critical juxtapositions to modern computing technology. They are highly site-specific, context-dependent and use simple, readily available, materials. The materials are mostly organic and prone to decay, alteration or fading. In the first two experiments, this plays a particularly integral part. The agency of the plant parts and the sand in "I Think I Have Lost You" change the materiality of the screen and its projected respectively projecting qualities. By choosing organic substances in "I Think I Have Lost You", I built a display from and with the surroundings. In their original function of providing an exchange of fresh air and sunlight between interior and exterior of the building, the windows are rendered useless. Their see-through quality is obscured, and they can no longer be opened. Instead, a new space is gained, and new phenomena occur. The sensations that the glazed-window displays produce, their double quality as screen and projector, is dependent on the time of the day and the weather condition. Transferred to the computer technology, this invokes questions of access and availability – of screens being always on(line). What if smartphone touchscreens were only available at certain times or in specific environments?[18]

In "Everything is Taken", the plant is both provider of the previously recorded and subsequent projected image and space for its projection. Most visitors of the gallery entered the space between the slide projector and the plant. In the beginning of the exhibition, it was often with this disruption that the projected image first became noticeable to the visitors because it illuminated their own bodies. The appearance of the plant changed during the exhibition time and the image became visible behind the installation. The screen quality of the plant was transferred to the background wall. The recognition of wall-as-screen proved to be much easier to recognize than that of the previous plant-as-screen. We are used to viewing flat, illuminated surfaces as screens. The installation invokes reflections on the time-conditioned and transient character of screenness, as well as it defamiliarizes what we regard as a screen. In this experiment, the human body disrupts the installation and momentarily and unintentionally becomes a partial screen itself. This points at the involvement of the body in screen culture and raises the question of whether eventually everything can be screen.[19] In the installation, the physical plant is the part that changes notably. When it comes to threads, I identified as relevant for computer screens, the experiment initiates a reflection on doubling/mirroring and representation. The agency of the living plant, its perishability and the lasting image do not form a simple, dualistic, opposition. Instead, comparable to augmented reality scenarios, a new intra-active screen space is established.

Screens as enactment, their materiality and the role of embodiment are even more obvious in "Everywhere, Nowhere". Here, water sprinklers and visitors to the park are in constant movement. On warm days, the water sprinklers motivate people to enter the space they provide. The fluidity of water makes the promise of the *Alice Through the Looking-Glass* narrative – the passing through the screen – experienceable. There is, however, no wonderland, to be entered. Furthermore, perceiving the water jets of the sprinklers as a screen in the first place is tied to my heightened sensitivity towards screenness. While filming the water sprinklers in Vyšehradské park, I had a conversation with a young woman who played Pokémon Go. She was on the hunt for one of the rarer Pokémon. For those who play, the smartphone display visualizes data structures and augments surroundings. Regarding screenness, the overlay of computer-generated content and physical world the game produces is noteworthy. In the park, the woman saw Pokémon, where I saw reflections on water jets. We comported ourselves towards different kinds of surfaces as screens. There is no all-embracing answer to what kind of screens where there and which were not.

Earlier, I stated that the aim of my experiments was to tackle screens outside the logics of computing. As a reference point, basic principles of computing (for example, input–processing–output models; standardization, formalization, computability), however, were never really absent during my research. Yet, the art-based approach allowed for a reconfiguration of the materiality and the modality of research. Put very simply, the basic principles of computing motivate a certain kind of knowledge production. Knowledge about the world that is already standardized or formalized is much easier to realize by a rule-oriented, mathematical device. To sum up, rather than giving answers to my initial questions, the quality of the art-based approach lies in its power to produce an experimental field of non-standard ways of knowledge production in a technological field. Computing already works with experimental settings, trial and error and translations between material (hardware) and semiotics (software). Artistic research, however, meant changing how these modalities can be addressed. During the experiments, the focus was not on optimizing a process or building a functioning piece of technology. Instead, a reflective and transformative way of addressing current computer screens was achieved through artistic re-translation. Over and above, the art-based experiments altered my perception and awareness of how screens are enacted. A special quality was the heightened embodied sensitivity – perceiving, sensing, touching, moving – in the moment of the site-specific study or experiment. This is a form of knowledge that, in all its richness, only insufficiently translates to paper. It points not just at what we know, but also through what we can know something and how we can pass along and express knowledge. In other words, it brings the performativity of knowledge production to the fore.

Moreover, the experiments provide ample connecting points for a further discussion on screens and screenness. In conclusion, I want to mention one aspect. With the *Alice in Wonderland* narrative in computing, I have characterized representational, doubling/mirroring and simulative, diffractive qualities of computer screens. Reviewing the art-based experiments, leads me to shift the focus for future research from the metaphors derived from optics, towards the notion of the screen as a membrane.[20] Indispensable for this was experimenting with organic materiality as constitutive for screens. Regarding screens as membranes highlights their active role. Thus, the screen's own agency as semi-permeable threshold, as well as its interconnectivity to specific sites, bodies and contexts, can be addressed.

Notes

1 Already in 1995, Lupton expressed the impact of the computer screen upon her work-life as a knowledge worker: "I am face-to-face with my computer for far longer than I look into any human face" (1995, p. 97).

2 I use both terms, "screen" and "display", sometimes interchangeable. Screen has a more generic meaning and in my view, puts an emphasis on the screen itself, whereas display connotes content, something has to be shown.

3 Also note the use of light guns and cathode ray tubes in video game technology (Wilk 2013, pp. 216–217).

4 Besides touch, other multi-modal interaction modes, like voice recognition, become increasingly popular, too.

5 "I find that existing studies tend to take the analytic status of screens for granted and juxtapose them with a human user to theorize the relationship between the two" (Ziewitz, 2011, p. 203). Even for the more popular field of urban screens, Erkki Huhtamo, who with "screenology" proposes a media archaeological approach, shares Ziewitz' view and finds that, "in spite of their growing prominence, public screens remain peripheral when it comes to media scholarship. Cinema and television studies, as well as 'new media' research, have largely ignored them. Most scholars of audiovisuality seem prone to look toward segregated and 'interiorised' (both psychologically and physically) experiences" (Huhtamo, 2009, p. 15).

6 Amongst others, Ziewitz refers to the works of Latour (1991; 2005), Verran (2001), Callon (1986) and Law (2007; 2008) (Ziewitz, 2011, pp. 208–209).

7 Heidegger's involvement in the German Nazi party (NSDAP) makes him a problematic resource. In what way his writings are influenced by antisemitic thinking has been discussed controversially (cf. Habermas & McCumber, 1989; Neske, Kettering & Harris 1990). Using a phenomenological approach without referring to his influential works is hardly possible, however.

8 "Being-in is not a 'property' which Dasein sometimes has and sometimes does not have, and *without* which it could just *be* just as well as it could be with it. It is not the case that man 'is' and then has, by way of an extra, a relationship-of-Being towards the 'world' – a world with which he provides himself occasionally. Dasein is never 'proximally' an entity which is, so to speak, free from Being-in, but which sometimes has the inclination to take up a 'relationship' towards the world. Taking up relationships towards the world is possible only *because* Dasein, as Being-in-the-world, is as it is. This state of Being does not arise just because some entity is present-at-hand outside of Dasein and meets up with it. Such an entity can 'meet up with' Dasein only in so far as it can, of its own accord, show itself within a *world*" (Heidegger, 1962, p. 84).

9 In contrast to the real-life Alice Liddell. Whether the stories are really based on her, as well as Carroll's friendship to Alice Liddell is discussed controversially in Leach, 1999.

10 "The notion of *intra-action* (in contrast to the usual 'interaction', which presumes the prior existence of independent entities/relata) represents a profound conceptual shift. It is through specific agential intra-actions that the boundaries and properties of the 'components' of phenomena become determinate and that particular embodied concepts become meaningful" (Barad, 2003, p. 815).

11 In a wider sense, the computer as a mirror of humanness is analyzed, for example, by Turkle (1984; 1995). For a detailed discussion of the computer screen as a semiotic mirror, see Draude, 2017.
12 For the use of diffraction in feminist science studies, see also Barad, 2007.
13 In short, a computer-generated bottle of milk works because it is referenced to our being-in-the-world – it speaks to our embodied knowledge.
14 *Water Tower*, translucent resin and painted steel installation, 1998. Available at https://www.moma.org/collection/works/82016?locale=en (accessed 20 Feb. 2017).
15 See "Noticing Screens: Introducing Screenness" above.
16 With *Überraschungsgeneratoren* Rheinberger refers to a term Mahlon B. Hoagland and with "Maschinen zur Herstellung von Zukunft" to an expression of François Jacob's, both molecular biologists (Rheinberger, 2012, p. 7).
17 For an in-depth discussion on how artistic practice serves to re-organize our perception and ways of thinking and become literally a "strange tool", see Noë, 2015.
18 Weather conditions, certainly, pose a challenge for current computer screens. Bright sunlight produces glare on displays and leaves projections invisible. To adapt computer screens to the time of day, there exist apps that filter out blue portions of the display light. In experience, the consistent availability and functioning of mobile technology everywhere is far from being a given. Hard- and software choices, interoperability and uneven power distributions limit its access and use.
19 Pranav Mistry from MIT media lab's fluid interfaces group, built a wearable gestural interface prototype where skin serves as a touchscreen, see http://www.pranavmistry.com/projects/sixthsense (accessed 24 Feb. 2017).
20 I owe this finding to a discussion of my experiments at the EU COST Action New Materialism Training School, Prague University, August 2017. Feminist STS scholar Astrid Schrader, in particular, helped me to develop this notion.

References

Barad, K. (2003). Posthumanist Performativity: Toward an Understanding of How Matter Comes to Matter. *Signs: Journal of Women in Culture and Society, 28*(3), 801–831.

Barad, K. (2007). *Meeting the Universe Halfway: Quantum Physics and the Entanglement of Matter and Meaning.* Durham, NC, and London: Duke University Press.

Barrett, G., & Omote, R. (2010). Projected-Capacitive Touch Technology. *Information Display, 26*(3), 16–21.

Beetz, J. (2016). *Materiality and Subject in Marxism, (Post-)Structuralism, and Material Semiotics.* London: Palgrave Macmillan.

Bolt, B. (2006). A Non Standard Deviation: Handlability, Praxical Knowledge and Practice Led Research. *Speculation and Innovation: Applying Practice Led Research in the Creative Industries.* Available at https://scholar.google.com.au/scholar?oi=bibs&hl=en&cluster=12501443088599024132 (accessed 30 July 2019).

Borgdorff, H. (2006). The Debate on Research in the Arts. *Sensuous Knowledge: Focus on Artistic Research and Development, 2*, 7–28.

Borgdorff, H. (2013). Artistic Practice and Epistemic Things. In M. Schwab (Ed.), *Experimental Systems: Future Knowledge in Artistic Research* (pp. 112–120). Leuven: Leuven University Press.

Busch, K. (2009). Artistic Research and the Poetics of Knowledge. *Art and Research: A Journal of Ideas, Contexts and Methods, 2*(2), 1–7.

Callon, M. (1986). The Sociology of an Actor Network: The Case of the Electric Vehicle. In M. Callon, J. Law & A. Rip (Eds.). *Mapping the Dynamics of Science and Technology* (pp. 19–34). London: Macmillan.

Carroll, L. (1992). *Alice's Adventures in Wonderland* [1865] and *Through the Looking-Glass* [1871]. In id., *Alice in Wonderland.* Ware, Herts.: Wordsworth Classics.

Dewey, J. (1934). *Art as Experience.* New York: Minton, Balch & Company.

Donner, C., & Jensen, H. W. (2006). A Spectral BSSRDF for Shading Human Skin. In T. Akenine-Möller and W. Heidrich (Eds.), *Proceedings of the 17th Eurographics conference on Rendering Techniques (EGSR '06)* (pp. 409–417). Eurographics Association: Aire-la-Ville.

Draude, C. (2017). *Computing Bodies: Gender Codes and Anthropomorphic Design at the Human–Computer Interface*. Wiesbaden: Springer VS.

Enticknap, R. G., & Schuster, E. F. (1959). Sage Data System Considerations. *Transactions of the American Institute of Electrical Engineers, Part I: Communication and Electronics*, 77(6), 824–832.

Esposito, E. (2003). Fiktion und Virtualität. In S. Krämer (Ed.), *Medien – Computer – Realität: Wirklichkeitsvorstellungen und Neue Medien* (pp. 269–296). Frankfurt am Main: Suhrkamp.

Habermas, J., & McCumber, J. (1989). Work and Weltanschauung: The Heidegger Controversy from a German Perspective. *Critical Inquiry, 15*(2), 431–456.

Haraway, D. J. (1992). The Promises of Monsters: A Regenerative Politics for Inappropriate/d Others. In L. Grossberg, C. Nelson & P. Treichler (Eds.), *Cultural Studies* (pp. 295–337). New York: Routledge.

Heidegger, M. (1962). *Being and Time* [1927], trans. J. Macquarrie & E. Robinson. Oxford: Basil Blackwell.

Huhtamo, E. (2004). Elements of Screenology: Toward an Archaeology of the Screen. *ICONICS: International Studies of the Modern Image, 7*, 31–82.

Huhtamo, H. (2009). Messages on the Wall: An Archaeology of Public Media Displays. In S. McQuire, M. Martin & S. Niederer (Eds.), *Urban Screens Reader, INC Reader #5* (pp. 15–28). Amsterdam: Institute of Network Cultures.

Introna, L. D., & Ilharco, F. M. (2006). On the Meaning of Screens: Towards a Phenomenological Account of Screenness. *Human Studies, 29*, 57–76.

Klein, J. (2011). Was ist künstlerische Forschung? *KunstTexte, 2*. Available at www.kunsttexte.de (accessed 30 July 2019).

Krajina, Z. (2009). Exploring Urban Screens. *Culture Unbound 1*(1), 401–430.

Krämer, S. (2011). Simulation und Erkenntnis: Über die Rolle computergenerierter Simulationen in den Wissenschaften. *Nova Acta Leopoldina, 377*(110), 303–322.

Kuhn, A. (2009). Screen and Screen Theorizing Today. *Screen, 50*(1), 1–12.

Latour, B. (1991). Technology is Society Made Durable. In J. Law (Ed.), *A Sociology of Monsters: Essays on Power, Technology and Domination* (pp. 103–131). London: Routledge.

Latour, B. (1993). *We Have Never Been Modern*. Cambridge, MA: Harvard University Press.

Latour, B. (2005). *Reassembling the Social: An Introduction to Actor-Network-Theory*. Oxford: Oxford University Press.

Law, J. (2007). Actor Network Theory and Material Semiotics (version of 25 April 2007). Available at http://www.heterogeneities.net/publications/Law2007ANTandMaterialSemiotics.pdf (accessed 20 Aug. 2018).

Law, J. (2008). On Sociology and STS. *Sociological Review, 56*(4), 623–649.

Leach, K. (1999). *In the Shadow of the Dreamchild: A New Understanding of Lewis Carroll*. London: Peter Owen.

Lewis, C. S. (2000). *The Complete Chronicles of Narnia* [1956]. New York: HarperCollins.

Lupton, D. (1995). The Embodied Computer/User. In M. Featherstone & R. Burrows (Eds.), *Cyberspace / Cyberbodies / Cyberpunk: Cultures of Technological Embodiment* (pp. 97–112). London: Sage Publications.

Manovich, L. (1995). An Archeology of a Computer Screen. Available at http://manovich.net/content/04-projects/011-archeology-of-a-computer-screen/09_article_1995.pdf (accessed 29 Jan. 2017).

McQuire, S., Martin, M., & Niederer, S. (Eds.) (2009). *Urban Screens Reader: INC Reader #5*. Amsterdam: Institute of Network Cultures.

Mol, A. (2002). *The Body Multiple: Ontology in Medical Practice*. Durham, NC: Duke University Press.

Nadin, M. (2007). Semiotic Machines. *Public Journal of Semiotics, 1*(1), 85–114.
Nake, F. (Ed.) (1993). *Die erträgliche Leichtigkeit der Zeichen: Ästhetik, Semiotik, Informatik*. Baden-Baden: Agis-Verlag.
Neske, G., Kettering, E., Harris, K. (1990). *Martin Heidegger and National Socialism: Question and Answers*. New York: Paragon.
Niantic (2016). Pokémon Go [mobile game]. Available at http://www.pokemongo.com (accessed 30 June 2018).
Noë, A. (2015). *Strange Tools: Art and Human Nature*. New York: Hill & Wang.
Rheinberger, H. R. (1992). *Experiment, Differenz, Schrift – Zur Geschichte epistemischer Dinge*. Marburg an der Lahn: Basilisken-Presse.
Rheinberger, H. R. (2012). Experiment, Forschung, Kunst. Vortrag, Jahreskonferenz der Dramaturgischen Gesellschaft, Oldenburg, 26.–29.04.2012. Available at http://www.dramaturgische-gesellschaft.de/assets/Uploads/ContentElements/Attachments/Hans-Joerg-Rheinberger-Experiment-Forschung-Kunst.pdf (accessed 30 June 2018).
Schwab, M. (2008). Convoluted Statement. Available at http://www.seriate.net/MSchwab_Convoluted-Statement_2008.pdf (accessed 30 June 2018).
Smithsonian National Museum of Natural History (2015), Bone Hall. Available at http://naturalhistory.si.edu/exhibits/bone-hall (accessed 30 June 2018).
Sutherland, I. E. (1965). The Ultimate Display. In W. Kalenich (Ed.), *Proceedings of the International Federation for Information Processing (IFIP) Congress* (pp. 506–508). London: Macmillan & Co.
Turkle, S. (1984). *The Second Self: Computers and the Human Spirit*. New York: Simon & Schuster.
Turkle, S. (1995). *Life on the Screen: Identity in the Age of the Internet*. New York: Simon & Schuster.
Verran, H. (2001). *Science and an African Logic*. Chicago: University of Chicago Press.
Vasseleu, C. (2013). Material-Character Animation: Experiments in Life-Like Translucency. In E. Barrett & B. Bolt (Eds.), *Carnal Knowledge: Towards a "New Materialism" through the Arts* (pp. 155–169). London and New York: I. B. Tauris.
Wilk, S. R. (2013). *How the Ray Gun got Its Zap: Odd Excursions into Optics*. New York: Oxford University Press.
Ziewitz, M. (2011). How to Attend to Screens? Technology, Ontology and Precarious Enactments. *Encounters* 4(2), 203–228.

12 Material Knowledge and Alchemical Practice

Katharina Vones

Introduction: Alchemy as a Meditation on Materials

The concept of alchemy continues to fascinate practitioners of contemporary craft who seek to create novel methodological approaches in order to address the increasing blurring of boundaries between artistic practice and scientific research. The figure of the alchemist has become a mythical conduit through which these practitioners can imagine themselves as interacting with complex technological and material solutions, while placing the artefacts they create as a result at the centre of the narrative. Forays into material experimentation and development by creative practitioners, such as Jolan van der Wiel and Helen Storey, offer new perspectives to both makers and researchers. The emergence of terms such as "creative technologist" (Connor, 2016), "craft technologist" (Shorter, 2017) and "technical creative" (Miodownik, 2003) to describe makers who operate at the boundaries of their disciplines and embrace collaborative methodologies hints at these developments. They circumscribe the complex interactions between maker and material that take place in the development of the technologically and materially sophisticated objects that manage, regulate and enrich nearly all aspects of our daily lives. In practice, the three terms introduced above lack clarity, and their meanings often remain obscured, as only a few practitioners engage with their proliferation in the field and too many alternative definitions exist.

This chapter examines the way in which the ancient practice of alchemy and the figure of the alchemist could be used to offer researchers and practitioners – in particular those operating at the boundaries between creative and scientific practice – a model for engaging with the concept of cross-disciplinary knowledge generation. Alchemical practice has been connected to craft practitioners, and in particular goldsmiths and jewellers, from the early modern period onwards (Principe, 2014). It experienced a resurgence within the field of studio jewellery through the epistemological changes that took place during the New Jewellery movement in the 1970s, where the altered perception of material preciousness in jewellery prompted some craft practitioners to return to the conceptual paradigms of alchemy in order to define their practice. More recently, the term "alchemical craft" has been used to describe practices and practitioners who work with novel materials and processes that have been sourced from the laboratories of researchers, often through interdisciplinary collaborative projects supported by an institutional framework. Their collaborators often come from the disciplines of physical and chemical engineering, material science, physics and biochemistry, amongst others, exploring experimental fabrication technologies and materials to venture into unforeseen territories where jewellery transcends traditional definitions of bodily adornment. Smart materials are an important part of this discussion, as their relative scarcity and transformative properties place them in the realm of both scientific research and artistic practice. Their status as materials with both functional and aesthetic properties has meant that while their use in a utilitarian context in disciplines such as medicine, engineering and architecture has become widespread, only a few artists have engaged on a deeper level with exploiting their unique characteristics. Materials libraries now exist globally as places where materials, including the unusual and extraordinary, are collected, catalogued and made accessible. They act as modern-day alchemical laboratories, where interested artistic practitioners, makers, materials scientists and academic researchers gather to experience and discuss novel materiality. Thus, a tradition of spaces for experimentation and the establishment of networks of practitioners, such as those secretive but well-documented meetings that took place between like-minded alchemists in sixteenth-century Europe (Principe, 2014), is revived at a time when such interdisciplinary collaborations are encouraged. By following historical tradition, they might offer a way to enhance the possible creative synergies between materiality, science and technology in order to provide an alternative path to cross-disciplinary knowledge generation.

This discussion will also examine how material knowledge can be accumulated and shared through a rigorous process of practical experimentation. Returning to the idea of vernacular knowledge generation through the application of experimental making practices, parallels are drawn between the alchemists of the early modern period and the alchemical jeweller in contemporary creative practice. In this respect, the connection between physicality and experiential knowledge generation suggested by Pamela Smith (2004) is of paramount importance. Smith closely examines how, in European traditional artisanal practice, the body was used as a vehicle for material knowledge generation and transference between creative practitioners, through following a methodology of close observation and repetitive labour. She states that:

> Artisans engaged in a bodily struggle with and against matter itself. Matter was not dead but alive, and it behaved in idiosyncratic ways, which artisans had to come to know – and master – through experience. (Smith, 2004, p. 114)

This statement echoes the meditative practice of the unknown craftsman advocated by Japanese philosopher and founder of the *mingei* folk art movement Soetsu Yanagi (Yanagi, Hamada & Leach, 1972), and parallels can be drawn between those two analogous paradigms. In both European and Japanese traditional artisanal practice, the craftsperson's knowledge of and profound connection to nature, emerges as central to advancing material knowledge. Yanagi explains that:

> Zen Buddhism uses the phrase *kenshō*, in which *ken* means "seeing" and *shō* "nature"; taken together, however, the two words do not mean "seeing nature" but rather "seeing into one's nature". In *kenshō* the artist and his guest are not two distinguishable concepts. (Yanagi et al., 1972, p. 152)

It becomes clear then that Yanagi sees the meditative practice of the craftsperson as a way to achieve oneness with the nature of all things by uncovering their essential characteristics. In European alchemical practice, pictorial and sculptural representations that imitate the essential processes of nature, such as depictions of the fly as a harbinger of putrefaction and regeneration, serve as symbolic metaphors to describe complex alchemical processes (Smith, 2004). As Smith states:

> Nature had to be imitated to bring into operation natural processes; processes that were bound up with the workings of the human body. In the artisan's attempt to imitate nature, he strove to create effects by employing the powers that inhered in nature. (Smith, 2004, p. 119)

By extension, alchemy can thus be regarded as a physical meditation on materials, with the expressed aim being the discovery of (and ultimate ability to replicate) their essential characteristics within alchemical practice. Experiential knowledge generation through the pursuit of an experimental practice that distils and thus reveals the essential nature of a material is an approach that could form the methodological basis for the creation of novel material knowledge directly relevant to contemporary craft practice. Just as the dichotomy between the natural and the artificial dominated the discourse of alchemy and the scientific method (Bensaude-Vincent, 2007), the polar opposition of art and science that took place throughout the twentieth century is being amalgamated again. As artists and designers increasingly engage with the structural complexities of the natural and material worlds in a more functional context, for instance in architecture and the emerging disciplines of bio art and bio design, scientists who previously primarily focused on functionality are striving to learn the secrets of harnessing creatively chaotic processes such as growth and spontaneous self-assembly from nature (Bensaude-Vincent, 2011).

In this context, particularly when reflecting on the place that smart materials[1] might occupy within this rhetoric, it is useful to point out the connection between alchemical practice and biomimetics. Biomimetics as an area of research looks to solve complex human problems through the close observation, analysis and imitation of models, structures and elements found in nature (Vincent et al., 2006). It has become an increasingly important methodological approach for contemporary art and design practice in the recent past. In a materials context, the principles of biomimetics are often applied to address issues of sustainability and energy conservation, either through using smart materials that exhibit biomimetic characteristics such as structural colour or hydrophobia or as part of complex architectural constructs that resemble biological organisms (Bhushan, 2009; Lloyd, 2003). Some creative practitioners utilise both of these characteristics, such as architect Doris Kim Sung in her installation "Bloom", which is made of a shape-changing bimetal and "breathes" when exposed to warming sunlight (Sung, 2012). The connection between alchemical practice and biomimetics is a profound one, with an early example of its theoretical application being Leonardo da Vinci's studies of birds' wings used as a basis for developing "flying machines" (Bhushan, 2009; Nychka & Chen, 2012). Pamela Smith concludes that:

> Four claims in particular emerge as common denominators [of vernacular knowledge generation]. First, nature is primary, and certain knowledge resides in nature. Second, matter is active, and one must struggle bodily with and against this active matter to extract knowledge of nature. Third, this process of struggle is called experience, and it is learned through replication. And, finally, this imitation of nature produces an effect – a work of art – that displays the artisan's knowledge

> of nature and in itself constitutes a kind of knowledge. The background to all these claims was the conviction that knowledge is active and knowing is doing. (Smith, 2004, p. 149)

These claims, their historical origins and how they relate to the contemporary creative practitioner, in particularly focusing on the emergent concept of the alchemical jeweller, will be examined in more detail in the following section.

Defining Alchemical Practice: The Traditional Goldsmith as Alchemist

The idea of the goldsmith as alchemist can be traced back as far as the early modern period, during which the ideas of *chrysopoeia* (metallic transmutation) and *chemiatria* (medicinal applications) as well as chemical production of enhanced materials crystallised into the practice of transmutational alchemy:

> Transmutational alchemy was about the transmutation of all base metals into more noble ones, but chrysopoeia was only one aspect of alchemy. Alchemy also touched on medicine and chemical manufacture. It was about the chemical production of things – medicines, porcelain, dyes, and other products as well as the precious metals – and about the knowledge of how to produce them. In this sense, "art technologies" – materials and techniques to make art and knowledge of these materials and techniques – overlapped with alchemy. (Dupré, 2014, p. xii)

These concepts and practical goals formed the basis for the emerging field of chymistry, which some believe to be a forerunner of modern chemistry (Principe, 2014). Chymistry proved to be an alluring field of experimentation for a broad spectrum of scholars, scientists, doctors and artisans, all of whom pursued a variety of shared goals related to material generation, knowledge and use. One of the most well-known of these quests was the development of the philosophers' stone, a type of elixir rumoured to be able to not just extract desirable components from metallic ores, but instead to effect a transmutation of base metals into precious ones. With their particular expertise in metallurgy assaying and workshops equipped with specialist metalworking tools as well as high-temperature furnaces, goldsmiths were perfectly positioned to participate in this endeavour. As Principe points out:

> Accordingly, gold- and silversmiths figure prominently in many examples of the genre of "transmutation histories" – detailed published accounts of successful transmutations – where such artisans are routinely called in to assay a sample of the gold or silver produced, and thus act as expert witnesses to its authenticity. (Principe, 2014, p. 160)

One particularly compelling example of a family of gold- and silversmiths who actively engaged in alchemical practice is that of Anthoni and Andries Grill. Of German origin, the two elder Grills, together with their younger brother Johannes, set up flourishing businesses in the Netherlands, making silverware and jewellery of high quality and great repute (Principe, 2014). While Andries' efforts in The Hague concentrated mostly on optimising precious metal extraction from Norwegian lead ore, Anthoni's ambitions were on a far grander scale, cemented by the establishment of six large laboratory spaces around Amsterdam for conducting alchemical experiments day and night. Both of the brothers engaged in well-documented collaborations with scientists, craftsmen and scholars, with their connections spanning several countries and social strata. The establishment of alchemical laboratories at European courts, such as the Uffizi gallery in Florence, served as spaces for intense interaction between practicing artists and scientists and helped to establish destinations for scientists to visit and engage in knowledge exchange with artisans (Kieffer, 2014):

> In these workplaces, a sort of hybrid figure was at work, with one foot in artisanal culture and another in scholarly culture and impossible to categorize in mutually exclusive categories of the scholar and the craftsman. Certain types of crafts – glassmaking, gold- and silversmithing, and porcelain production – seem to have been particularly prone to exchanges between artisanal and scholarly alchemical cultures. (Dupré, 2014, p. xv)

Networks of alchemical practitioners began to emerge in many of the larger cultural centres, such as London, Paris and Amsterdam, spreading their recipes and speculative practices amongst a tight-knit and somewhat secretive circle in the form of writings, recipes and personal accounts (Dupré, 2014). The dissemination of these materials as manuscripts, texts and books and the presence of these in the artisanal workshops denoted a cultural shift – where formerly skills had only been passed on from master to apprentice by means of demonstration and practice, now scholarly learning had entered the realm of the workshop. With the arrival of the scientific method favouring quantitative experimentation and scientific rigour over lore, the practice of alchemy experienced a steady decline, and by the twentieth century only the mythical figure of the alchemist remained. As a consequence, the division between skill-based artisanal crafts, fine arts and scientific research had become all but absolute.

The Contemporary Jeweller as Alchemist

With the emergence of the New Jewellery movement in 1970s Europe, which occurred due to a rising engagement and exchange of ideas between the Dutch, German and British studio jewellery communities, the idea of the jewellery artist as alchemist was resurrected. At the heart of this movement, that in time came to critically shape the perception of contemporary jewellery, lay the concept of questioning commonly accepted ideas of preciousness by experimenting with unorthodox materials, engaging with social issues and creating a new awareness of the body and the wearer (Skinner, 2013). Thus an epistemological change occurred – a move for the alchemical concept of *chrysopoeia*, to signify the creation of preciousness in contemporary jewellery and objects through the touch of the jewellery artist's hands rather than by using materials of inherent value. This epistemological transformation also aids in aligning the New Jewellery movement with similar developments that took place in the fine arts throughout the twentieth century, in particular elevating the status of the artisanal crafts practitioner from mere maker to that of artist and transformative auteur. As Susan Cohn points out:

> Makers became influenced by debates and movements across art, craft and design, and were impacted by the rise of new consumer and media cultures. Together, they sought to make jewellery democratic and renew its connection to the body, to question jewellery's social role and experiment with new materials. (Cohn, 2012, p. 224)

Jewellery artists such as Ruudt Peters have continued to engage with these ideas through exploring concepts of spirituality, materiality and the role of the maker in shaping altered perceptions of value. By creating body ornamentation, both figurative and abstract, that he considers to be in the tradition of the transmutational alchemist (Bernabei, 2011), Peters attempts to transform materials of little monetary value into precious objects imbued with personal meaning. For Peters "alchemy is not an exact science. It is a 'proto-science', in which the intuitive prevails over the rational" (see Baas, 2015). These spontaneous and sometimes anarchic material transmutations, particularly evident in Peters' work from the 1990s onwards starting with the Ouroboros series (Peters, 2015), continue to influence the way materials are viewed and used by practitioners in the field of contemporary jewellery. Through the act of obscuring and revealing layers of different substances, such as can be seen in his pieces from the Iosis and Azoth series, Peters engages in a constant dialogue of material exploration. The titles of these experimental series point towards alchemical traditions, while the materials themselves are inherently modern and often synthetic, including liquid polyester resin, dyes and foam. As Love Jönsson points out:

> There are links between the materials' metamorphosis and the artist's own development. The creation and recreation of his identity has been a recurring point of reference in Ruudt Peters' work. Once again the threads lead to alchemy where the transmutation of the material has the function of a portrayal or a projection of the alchemist's spiritual development. (Jönsson, 2004)

As is often the case with contemporary jewellery, the presentation of the work in exhibition forms an essential part of its interpretation. In the case of the Ouroboros series, its installation, high on the wooden ceiling of Gallery Marzee, visible only by climbing precarious wooden ladders, contributed significantly to its connection with the concept of alchemy. As Liesbeth den Besten recounts:

> The use of this device, the ladder, was not just a rhetorical "trick": in alchemy – the main source for Ruudt Peter's work – the ladder is a symbol that represents reaching for the divine. The public was invited to climb the dangerous, old-fashioned ladders that stood in rows under the beams of the ceiling in order to get a closer view of the jewellery – indeed to see the jewellery at all, as nothing could be seen from ground level. (den Besten, 2011, p. 52)

For Peters, the curated experience of viewing his jewellery forms an essential part of both its concept and communicated meaning. Rooted in the fine art traditions of installation art and "happenings" from the 1960s onwards (den Besten, 2011), Peter's artistic background as a sculptor fostered a greater appreciation for the planned performance in his practice as a vehicle for artistic intent that to this day continues to influence the way in which contemporary jewellery is displayed:

> Well, after my jewellery studies I worked for ten years as a sculptor, and that made the space and the surroundings very important. I curate each solo exhibition by the millimetre, and the installation has to be more or less a certain kind of finger pointing which directs the audience to the soul behind my work. (Bernabei, 2011, p. 157)

Artists such as Lisa Walker and Karl Fritsch, on the other hand, have reinterpreted the challenge to critique preciousness through material use by adapting their work processes to subvert the traditional methods and conventions of jewellery-making. Walker in particular engages with the way in which the process of using glue has long been frowned upon by craftsmen, through creating colourful pieces that use glue as an essential structural and aesthetic component. She does not as such distinguish between precious and non-precious materials – for Walker, the materials that regularly surround us in a society of abundance are given new meaning through the transformative touch of the maker, challenging the wearer through a subversion of material and context: "I don't want to make pieces that are easily steered through our established channels, I want people to be forced to work on new syllogisms, analogies and positions" (Walker, 2015).

However, the last two decades of contemporary jewellery have seen little further development of these concepts, and critics like Susan Cohn argue that:

> By the 1990s, signature had been invented and repetition settled in. For some, the Contemporary Jewellery Movement had not just reached its zenith but come to a decisive end – even as contemporary jewellery continues unabated as an inventive practice ... increasing emphasis has been placed on authentic stories in craft and "star" status of the maker, over renewed questioning about jewellery values. (Cohn, 2012, p. 231)

Such criticisms are justified and point to the necessity for the discipline to re-invent itself in the face of the increasing popularity of branded luxury jewellery and the passing of the generational baton amongst practitioners. New practitioners are subjected to the inexorable political machinations of a gallery system that prioritises commercial viability over conceptual radicalism outside an established framework of semi-sanctioned subversion, thus leaving little room for divergent thought. This is a line of criticism that could be levied equally at most of the contemporary art world, and has to some extent been tempered recently by the rise of democratised making through the FabLab and Hacker movements (Gershenfeld, 2012). For the studio jeweller, a rising interest in collaborative projects between the arts and sciences offers not only the possibilities of exploring novel technologies and materials, but also signals a chance to reclaim the term "alchemical jeweller" to be more aligned with its historical origins. Responding to future challenges in diverse areas of research that are not directly related to the visual arts could provide the starting point for such a transformation.

Artistic Practice as Transmutative Force

This chapter has explored the profound connections that exist between artistic and alchemical practice, particularly if viewed in relation to the field of contemporary jewellery. A generation of jewellery artists has emerged that revel in exploring cutting-edge technologies and materials, pushing the boundaries of what constitutes jewellery in the process. Often the most successful and conceptually strong work is the result of artists collaborating with scientific partners, thus resurrecting the spirit of knowledge exchange and mutual growth evident in the original alchemical laboratories at European courts (Kieffer, 2014). One such example is the nano-particle infused, scented jewellery, conceived by Sofie Boons in conjunction with research scientist Jodie Melbourne. A self-proclaimed "alchemical jeweller" (Boons, 2013), Boons aims to make the invisible visible and continues to explore novel ways to interact with technologically advanced smart materials in order to investigate the conceptual relationship between scent and gold (Atherton, 2015). However, Boons' is a very straightforward approach to collaborative knowledge generation that does not necessarily seek to question its conceptual underpinnings nor deeply investigate its relationship with novel materiality.

As the ideology of critical design, first suggested by Anthony Dunne (1999), has gained traction as a methodological approach, a growing number of jewellery practitioners have engaged with its underlying principles of using speculative design proposals as a tool to question assumptions about the role of the object within an increasingly technologically and materially sophisticated society. Examples of this provocative approach are the speculative jewellery prototypes created by Israeli industrial designer Naomi Kizhner, that hint at a future in which the body could be used for energy harvesting on a small scale (Treggiden, 2014). Designed to intersect with the veins and nerve pathways of the spinal cord, Kizhner's Energy Addicts are made of gold and 3D-printed biopolymer and use bodily fluids and impulses to operate miniature mechanical structures that in turn generate miniscule amounts of energy. Kizhner's prototypes evoke the idea of boundary transgression – between the mechanical and the biological, the inner workings of the human body made visible by a wearable object that situates itself both inside and outside of its natural confines at the same time. Another example of the application of critical design in a jewellery context is the Biojewellery Project led by jewellery designers Tobie Kerridge and Nikki Stott in conjunction with bioengineer Ian Thompson (Thompson et al., 2006). Through harvesting dental tissue from the mouths of engaged couples, and growing it around a ring-shaped scaffold with the help of a bioengineer, wedding rings made of bone tissue belonging to the prospective spouse were created. This might seem like the ultimate tokenistic gesture of commitment in an age where the human body is increasingly viewed as a site for liberal surgical intervention in the name of functional and aesthetic enhancement. However, the notion of a wearable object made from another's bodily tissue remains deeply uncanny, despite well-known instances of this practice in, for instance, Victorian mourning jewellery made from the hair of the deceased (Lutz, 2011). It is the procedural intervention of the bioengineer that turns Thompson and colleagues' work into a powerful and provocative critique of the idea of the symbolic value of adornment. By harvesting living proto-cells and manipulating them into a designed, predetermined shape, the natural is turned into the artificial and thus undergoes the process of boundary transgression between living and dead that is associated with the uncanny object.

The concept of the uncanny object is also a defining paradigm of my own practice as a jewellery artist. Working mainly with silicone that has been mutated into a smart, proto-organic material through the addition of thermochromic and photochromic pigments (Plate 13), the resulting objects are biomimetic and react autonomously to changes in temperature and light when worn on the body. The uncanny connotations of creating an artificial yet living organism are emphasised by the perception of synthetic polymers as "chameleon materials" whose inherent plasticity and unlimited potential for functional adaptability and aesthetic imitation stands in stark contrast to the solidity and predictability of more traditional materials (Bensaude-Vincent, 2007). In my making practice I apply a hybrid methodology that draws on the alchemical process of vernacular knowledge generation, by effecting a deep engagement with the material through practical experimentation, while applying a methodical approach to accurately recording the resulting findings, thus ensuring repeatability. The personal materials library of physical samples that represents the outcome of this process (Plate 14) can thus be used to communicate these findings efficiently and accurately to other creative practitioners. The combination of scientific recording processes and vernacular knowledge generation through the direct, tactile interaction with materials is a novel methodological approach in the field of studio jewellery.

The aim to refine and deepen the awareness creative practitioners have of the usefulness of these recording processes and test protocols in the context of their own practice through collaborations with scientists, technologists and engineers is at the heart of the alchemical practice. Encouraging collaborative research endeavours between artistic and scientific practitioners is essential in establishing a meaningful methodological dialogue between science and technology studies and the contemporary crafts. These collaborations still face many obstacles – not least, as Kerridge notes:

> This disconnection between the present and our ideas about the future is largely due to a lack of dialogue at the stage of design and development, between those that create technology and those that use it. We may no longer imagine men in white coats inventing the future – a vision propagated in the 1950s – but it still comes down to "them" and "us". (Thompson et al., 2006, p. 9)

Miodownick's "technical creative" is a type of practitioner who is well-versed in both scientific and artistic methodologies, equipped to contribute to the debate surrounding the role of the creator in an age defined by digital revolution and material discovery (Miodownik, 2003; Wilkes et al., 2016). In his advocacy of enhanced knowledge creation through a deeper cultural engagement on the part of scientists, Miodownik harks back to the alchemical practice of the early modern period. Before the strict division of experiential knowledge, created through the vernacular process of artisanal practice, and the emergence of modern science as a discipline, focused on the intellectually elevated knowledge gained through the philosophical formulation of theory, took place (Smith, 2004), the collaborative practice of the alchemist was an essential connection between these disparate views. As Pamela Smith points out:

> Discussions of artistic and scientific innovation have traditionally focused on the individual: the artistic genius, or the lone scientist. This elevation of the individual genius to centre stage ... has often formed the core of a narrative of modernity and progress. Recent scholarship in the areas of distributed cognition and indigenous or vernacular knowledge systems, however, has begun to portray the production of knowledge as the result of a complex and locally situated relationship among many individuals and groups at different levels of society. (Smith, 2004, p. 240)

Conclusion

It becomes clear, then, that a separation of the arts (including craft and design) and the sciences – of practice-led, experiential knowledge and scientific theory based on methodical protocol – can only hamper the progression of both This is particularly true in the context of the emerging importance of a constantly changing, all-encompassing materiality that requires scientific insight and rigour, as well as creative sensitivity to material quality in order to be utilised to its fullest potential. This chapter represents an attempt to summarise these developments, particularly if viewed in conjunction with the increasingly fruitful practice of biomimetics. As mentioned in the introduction, both biomimetics and alchemical practice seek to extract the knowledge residing within nature through a process of profound analysis and interpretation. However, whereas the practice of biomimetics relies mostly on processes of close observation and imitation to achieve this aim (including through the use of technology), alchemical practice depends on the tactile and experiential vernacular knowledge created through direct experimental interactions with materials and the meticulous recording of the resulting knowledge for the purpose of communicating the findings to other practitioners. Through the application of either or both methodologies, artistic practice thus holds the potential to act as a transmutative force, particularly if applied as a lens through which the 'traditional' natural sciences and their hermeneutic modes of knowledge creation are viewed. The conceptual return to alchemy during the nascence of nuclear science as a scientific endeavour affirms the deep-seated human need to turn to a more spiritual, creative approach to knowledge creation if faced with hitherto inexplicable natural phenomena. As Morrisson points out:

> Even religious skeptics began to wonder if the alchemists might have understood something about the nature of matter that nineteenth-century scientists had missed. … But the new science of radioactivity had superseded Dalton and began to suggest, even to scientists, that the bases of alchemy and those of the new chemistry were not mutually exclusive. (Morrisson, 2007, p. 11)

It is no coincidence that the terms used to describe creative practitioners mentioned in the introduction – "technical creative", "craft technologist" and more recently "alchemical jeweller" – have focused on an amalgamation of different modes of knowledge creation. United in fruitful collaboration, they need to be at the centre of future creative and scientific practice in order to achieve a transmutation of their respective qualities into a sum greater than its respective parts.

Note

1 Smart materials are designed materials that respond to externally applied stimuli, such as pressure, stress, electrical and magnetic impulses as well as changes in temperature, moisture, pH and light levels with a controllable and often reversible state change.

References

Atherton, G. (2015). Design Life: Sofie Boons. Retrieved from http://www.material-lab.co.uk/blog/post/design-life-sofie-boons.

Baas, F. (2015). Ruudt Peters: Statement. Retrieved from http://klimt02.net/jewellers/ruudt-peters.

Bensaude-Vincent, B. (2007). Reconfiguring Nature through Syntheses: From Plastics to Biomimetics. In B. N. Bensaude-Vincent & R. William (Ed.), *The Artificial and the Natural: An Evolving Polarity* (pp. 293–312). Cambridge, MA: MIT Press.

Bensaude-Vincent, B. (2011). A Cultural Perspective on Biomimetics. In A. George (Ed.), *Advances in Biomimetics*, IntechOpen. DOI: 10.5772/10546.

Bernabei, R. (2011). *Contemporary Jewellers:Interviews with European Artists*. London: Bloomsbury.

Bhushan, B. (2009). Biomimetics: Lessons from Nature: An Overview. *Philosophical Transactions of the Royal Society A, 367*, 1445–1486.

Boons, S. (2013). Sofie Boons Artist: Homepage. Retrieved from http://sofieboons.com.

Cohn, S. (2012). A Brief Contemporary Jewellery History. In S. Cohn (Ed.), *Unexpected Pleasure: The Art and Design of Contemporary Jewellery*. Melbourne and London: Skira Rizzoli and Design Museum.

Connor, A. M. (2016). A Historical Review of Creative Technologies. In A. M. Connor & S. Marks (Eds.), *Creative Technologies for Multidisciplinary Applications*. Hershey, PA: Information Science Reference (IGI Global).

Den Besten, L. (2011). *On Jewellery: A Compendium of International Contemporary Art Jewellery*. Stuttgart: Arnoldsche Art Publishers.

Dunne, A. (1999). *Hertzian Tales: Electronic Products, Aesthetic Experience and Critical Design*. London: Art Books International Ltd.

Dupré, S. (2014). Introduction. *Archimedes: New Studies in the History and Philosphy of Science and Technology, 37*, i–xx.

Gershenfeld, N. (2012). How to Make Almost Anything: The Digital Fabrication Revolution. *Foreign Affairs, 9*(6), 43–57.

Jönsson, L. (2004). Stone and Water: Notes on Ruudt Peters' Azoth. *Ornamentum*. Retrieved from http://www.ornamentumgallery.com/artists/ruudt-peters/series/azoth-2004.

Kieffer, F. (2014). The Laboratories of Art and Alchemy at the Uffizi Gallery in Renaissance Florence: Some Material Aspects. *Archimedes: New Studies in the History and Philosphy of Science and Technology, 37*, 105–127.

Lloyd, A. (2003). Biomimetic Supporting Materials: Shape Memory Alloys. *Materials Today, 6*(1), 14. DOI: 10.1016/S1369–7021(03)00117-2.

Lutz, D. (2011). The Dead Still Among us: Victorian Secular Relics, Hair Jewelry, and Death Culture. *Victorian Literature and Culture, 39*(1), 127–142. DOI: 10.1017/S1060150310000306.

Miodownik, M. (2003). The Case for Teaching the Arts. *Materials Today, 6*(12), 36–42.

Morrisson, M. (2007). *Modern Alchemy: Occultism and the Emergence of Atomic Theory*. Oxford: Oxford University Press.

Nychka, J. A., & Chen, P.-Y. (2012). Nature as Inspiration in Materials Science and Engineering. *JOM: The Journal of the Minerals, Metals and Materials Society, 64*(4), 446–448.

Peters, R. (2015). Ruudt Peters: Homepage. Retrieved from http://www.ruudtpeters.nl.

Principe, L. M. (2014). Goldsmiths and Chymists: The Activity of Artisans Within Alchemical Circles. *Archimedes: New Studies in the History and Philosophy of Science and Technology, 37*, 157–179.

Shorter, M. A. (2017). A Craft Technologist's Approach to Paper Electronics and Circuitry. (unpublished PhD thesis), University of Dundee.

Skinner, D. (2013). The Critique of Preciousness. In D. Skinner (Ed.), *Contemporary Jewellery in Perspective* (p. 63). New York: Lark Jewellery & Beading.

Smith, P. H. (2004). *The Body of the Artisan: Art and Experience in the Scientific Revolution*. Chicago and London: University of Chicago Press.

Sung, D. K. (2012). Metal That Breathes: TEDxUSC. *TED Talks*. Retrieved from https://www.ted.com/talks/doris_kim_sung_metal_that_breathes.

Thompson, I., Kerridge, T., & Stott, N. (2006). Biojewellery: Designing Rings with Bioengineered Bone and Tissue. Retrieved from http://research.gold.ac.uk/2317/2/biojewellery-booklet.pdf.

Treggiden, K. (2014). Naomi Kizhner's Jewellery Collection Harvests Energy from the Human Body. Retrieved from https://www.dezeen.com/2014/08/06/naomi-kizhner-energy-addicts-jewellery-human-electricity-production.

Vincent, J. F. V., Bogatyreva, O. A., Bogatyrev, N. R., Bowyer, A., & Pahl, A.-K. (2006). Biomimetics: Its Practice and Theory. *Journal of the Royal Society Interface, 3*, 471–482.

Walker, L. (2015). Lisa Walker Jewellery Artist: Homepage. Retrieved from http://www.lisawalker.de/home.html.

Wilkes, S., Wongsriruksa, S., Howes, P., Gamester, R., Witchel, H., Conreen, M., Laughlin, Z., & Miodownik, M. (2016). DesignTools for Interdisciplinary Translation of Material Experiences. *Materials and Design, 90*, 1228–1237.

Yanagi, S., Hamada, S., & Leach, B. (1972). *The Unknown Craftsman: A Japanese Insight into Beauty*. Tokyo: Kodansha International Ltd.

13 Kissing and Staring in Times of Neuro-mania

The Social Brain in Art-science Experiments

Flora Lysen

Introduction: Brains Buzzing in Art Space

Brains have entered the gallery. Over the past decades, measurements of the brain's electrical activity have moved beyond the neuroscientific laboratory into other domains, including practices of mindfulness training and meditation, hacker spaces, consumer research, the game industry and also a variety of art-meets-science events. Now that headsets for EEG (electro-encephalography) measurements and software for BCI (brain-computer interfaces) applications have become more affordable, colorful visualizations and tantalizing sonifications of the active brain have started to pervade science festivals, tech-innovation conferences and art institutes. Consider, for example, Stelarc's 2010 performance *Spectacle of Mind*, in which the well-known performance artist wore an EEG headset to create a 'brainwave-generated' art piece, a Surrealist-looking animation of his own face projected for an audience on a huge screen (Stelarc, 2010). Or artist Lisa Park's installation *Eunoia* (2017), in which EEG measurements cause buzzing noises and ripples on the surfaces of a number of water basins, depending on the levels of the performer's concentration. In these varying art-science installations, glowing lights, buzzing sounds, color-changing blobs and spiraling lines indicate that on a neural level, things are continuously changing. Often, these experimental mediations of brain activity are deliberatively incomprehensible: we do not know precisely what particular cognitive state, mood or behavioral pattern we are witnessing. Instead, viewers of these art-science works are asked to draw intuitive relations between the subjects and data-environments on display: that is, to see these elements as continuously affecting one another in an intricate feedback loop. Moreover, these performative installations ask viewers to pay special attention to a usually invisible element within this assemblage: the strange and striking real-time inscriptions of the brain's electrical activity.

Two recent art-science experiments with EEG particularly probe this relation between brains, bodies, technologies and the surrounding environment: the work *Mutual Wave Machine* (2013–present), by Suzanne Dikker and Matthias Oostrik and the work *EEG KISS* (2014–present) by Karen Lancel and Hermen Maat (Figure 13.1, Plate 15). The first art-science experiment *Mutual Wave Machine* invites participants with EEG headsets to examine the notion of 'being on the same wavelength', while the second uses EEG to visualize the brain activity of two people kissing in a public space, thus probing notions of intimacy and measurability. In this chapter, I analyze these two art-science experiments to demonstrate how both – in different ways – reconfigure and reconsider research into the 'social brain'. Current research into the

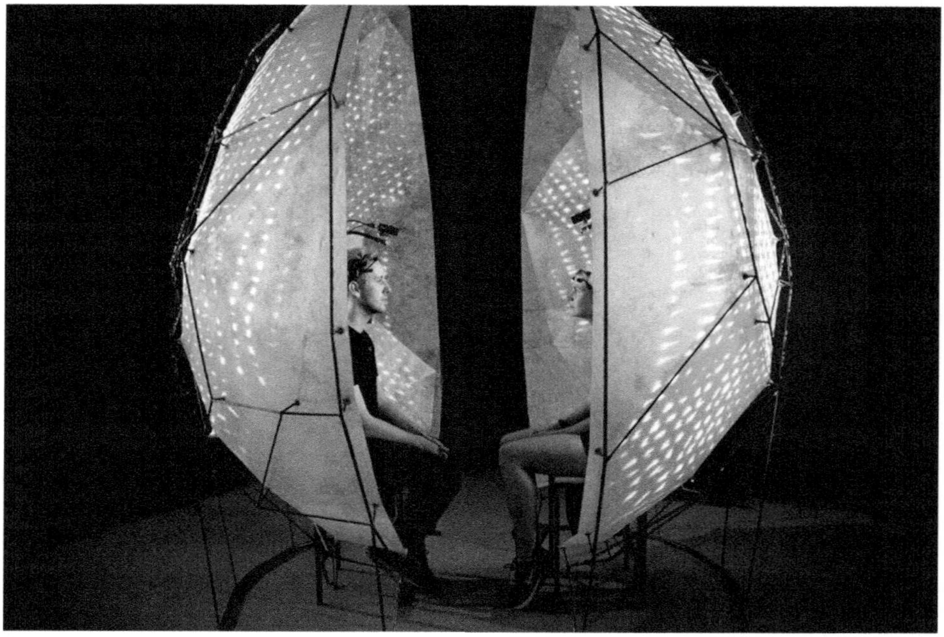

Figure 13.1 *Mutual Wave Machine* (Dikker & Oostrik, 2014). Image courtesy of Lexus Hybrid Art, 2014.

domain of brain science – and the social brain in particular – is, as I explain, a contentious area, in which art-science experiments can play an important investigative part. Building on recent work at the intersection of artistic research and STS, I investigate these art-science works as public experiments with potential elements of 'experimental entanglements': that is, as configurations of knowledge (developed with participants) with a special exposure of disciplinary boundaries and definitions (Fitzgerald & Callard, 2015, p. 18). Ultimately, I argue that such art-science installations, as entangled experiments, can help to reimagine the empirical and conceptual outlines of research into the social brain but that such reflections are also always paired with other logics and tendencies, including, for example, art's position in an innovation-oriented neuro-techno-scientific society and art's relation to 'neuro-enchantment'.

Art-Science Experiments Meet Science and Technology Studies

When speaking of 'art-science experiments' in this chapter, it is of course the hyphen between 'art' and 'science' that has been the central issue in discussing this field of artistic experimentation. The question is whether this conjunction of 'art-plus-science' constitutes truly shared research or, instead, whether the hyphen is evidence of the preconceived differences and hierarchies between the fields of 'science' (mostly used as a shorthand for 'natural sciences') and 'art' (Halpern, 2012; Lapworth, 2016; Michael, 2012; Wienroth & Goldschmidt, 2017; Zschocke, 2010). As various authors have noted, art–science collaborations – most notably those in the field of bio-art – often take the form of art that (critically) elucidates or examines scientific practices.

With this framing, as Rob Zwijnenberg notes, 'art's gift to science', is often characterized as an art that serves 'to help us (and science) understand or become critically aware of the implications of science or to help us (and science) reshape culture in the face of technological developments' (2012, p. 7). Understood as such, the role of art is chiefly instrumental: art functions as a moral compass in scientific development or as a special means to contribute to scientific innovation. Art–science collaborations, in this line of reasoning, are multidisciplinary or interdisciplinary (intermediating between the fields of art and science), yet they rarely constitute, as Henk Borgdorff argues, genuine artistic research in the sense of a 'real hybridization of domains' (2010, p. 53).

In this chapter, I do not presuppose nor dismiss the status of the hyphen in art-science practices. Instead, I follow several other studies in taking this dichotomy as itself produced by experimental boundary work that deserves investigation (Gieryn, 1983). As such, in this chapter, I build on recent intersections between science and technology studies (STS) and reflections on artistic research (Born & Barry, 2010; Calvert & Schyfter, 2016; Michael, 2012). The scholarly upshots of such intersections are multifold. Perhaps the most straightforward observation on the potency of interfacing STS with the field of art is that it entails an important focus on the *work* of art and the many agents necessary to assemble the work of art. Because STS focuses on 'science-in-the-making', joining it to the study of art would mean, as Ruth Benschop and others have emphasized, a conception of the art work as an 'ongoing endeavor of assembling agencies, rather than constructing a finished work that can be (re)presented in a more or less unproblematic way' (Benschop et al., 2014, p. 40). Important, in this scholarly frame, is that 'art', 'artistic', 'science' or 'research' are not predefined categories but emerge differently according to particular situations.

A recent publication in the *Handbook of Science and Technology Studies* extends these observations on the potential of interfacing STS and research through art and design. Chris Salter, Regula Burri and Joseph Dumit argue that strategies taken from research through the arts can be used as *methods* for STS inquiry and that art-science work can open up new directions and *perspectives* of research. When STS would be more open to research through the arts, 'STS itself becomes experimental; it not only observes people "thinking with eyes and hands" but uses eyes and hands to intervene and interfere in spaces and sites where science and technology are constructed, distributed, used, incorporated, and enacted.' (Salter et al., 2016, p. 154). The analysis of two art-science experiments in this chapter draws on these observations: as forms of STS-inspired artistic research, art-science experiments similarly combine analysis and practice, and the art-science work can be viewed as bringing new perspectives to the fields in which it partakes and also as bringing forth novel methods. It is in this sense that I will speak of the art-science work as enabling both a *reconsideration* and *reconfiguration* of experimental systems.

Particularly important for my discussion of art-science experimentation is the work of sociologists Georgina Born and Andrew Barry, who delineate three different and intersecting ways (what they call 'logics of interdisciplinarity') of understanding the relation between art and science in art-science practices (Born & Barry, 2010). First, Born and Barry point to the existence of a 'logic of accountability' (1) through which art-science works are used as means to render scientific knowledge more accountable to society, a way to stimulate debate and engagement with science. In this view art-science functions as an instrument to gather a public for science and in turn to communicate and educate about science for these publics. Secondly, the 'logic of innovation' (2) is in play when art-science is viewed as a partner in providing new insights for the development of new (commercial) products in an innovation-oriented economy. Born and Barry note how these two logics both render art subservient to the course of science and technology, engendering a hierarchical picture in which science is 'complete' and merely needs to be mediated, explicated and utilized. Alternatively, the authors are most in favor of a third 'logic of ontology' (3), a relationality that constitutes true 'hybridization' of fields: this is when art-science practices redefine the object of research and the subjects and publics engaging with it, contributing 'to the generation of something new within scientific practice itself, challenging the boundaries of disciplinary authority' (Born & Barry, 2010, p. 114).

Adding to Born and Barry's tripartite scheme, I propose art-science experiments can offer a fourth logic, what I call a 'logic of affect' (4): they open up ways for experts and non-experts to be newly affected through bodily participation in experimental systems. Public, participatory experiments such as *Mutual Wave Machine* and *EEG KISS* particularly conjure such affective dimensions, for example, when they invite audience members to become participants, to have an experiential dimension of being part of a scientific measuring situation and, in the case of EEG artworks, to have an experience of the body in relation to technical mediation of neural activity. This, as art historian Caroline Jones contends, 'is exactly what the artist can bring into being that the scientist does not: a way to make empirical research come alive in the body of the viewer' (2012, p. 169). With this affective dimension of art-science works I denote their ability to open up new possibilities for the body's capacity to act – 'the ability to affect and be affected' (Massumi, 2015, p. ix). Conversely, affective dimensions of art-science experiments may also tie in with a logics of accountability (1), when they create impressive experiences that make viewers more amenable to the desires, anticipations and excitement of science institutions (hence, art-science as a 'device for the governance of affect' (Born and Barry, 2010, p. 109; cf. Anderson, 2007). However, this affect-induced accountability does not preclude the possibility of affect in the sense of experiential insight. When visitors not only witness but partake themselves in a scientific experiment, art-science installations lend an affective dimension to the unfolding of empirical research. I proceed with analysis of two art-science experiments *Mutual Wave Machine* and *EEG KISS* with a focus on these (continuously intersecting) logics of affect, accountability, innovation and ontology. I first position these two works within a broader movement of 'critical neuroscience' and question the position of brains exhibited and performed beyond the scientific laboratory.

From Neuro-hype to Neuro-critique

Attending to current discourses on the position of brain science, one can start to see why art-science experiments are especially imperative in this academic arena and why potential reconfigurations of neuroscientific concepts stemming from these art-science experiments are so important today. Over the past decade, a number of scholars have become critical of the omnipresence of brain science in cultural discourses – sometimes dubbed 'neuro-hype' or 'neuro-mania' – and the increasing importance attributed to the brain to understand complex phenomena ranging from adolescent behavior and gender differences to consumption, criminality and depression (Choudhury & Slaby, 2011; Littlefield, 2017). In different ways, a scholarly movement of 'critical neuroscience' has cautioned against today's ubiquitous neuro-centrism: firstly, because it does not do justice to the embodied, enculturated, enactive and affective dimension of human behavior, and, secondly, because neuro-centrism may feed into what has been called 'neuro-governmentality': a brain science that is prescribing people how to live their lives; that builds norms and politics into brain facts; and that makes social things look nature-like.[1] Most discontent is directed towards the iconography and rhetoric of the neurosciences – the discipline's beautiful images and science-fictional promises – which may run ahead of actual development of the science and the technologies (Beaulieu, 2002; Meynell, 2012; Roskies, 2010). Colorful functional brain scans are at the forefront of this promissory culture: with their seductive reality effects and promise of objectivity they can cause the 'apparent materialization' of emotion, affect and cognition at the level of observed processes in the brain (Meloni, 2011, p. 310). Neuro-imaging technologies are part of a 'translational imperative' (an eagerness to translate and interpret laboratory findings as real-life behaviors, experiences and situations) and with this approach, as various ethnographers of neuroscience have argued, translational steps may quickly lead to a loss of complexity (and perhaps even 'reckless extrapolation') in understanding the phenomena under investigation (Rose & Abi-Rached, 2013a, p. 228).

Over the past two decades, these cautioning critiques have particularly been raised in the emerging field of SCAN sciences (the social, cognitive and affective neurosciences), and the development of a new understanding of a 'social brain'. Through the SCAN sciences (a wide-ranging and emergent field with much internal contestation and heterogeneity), a new concept of the brain has been called into being, a brain that supports and predisposes our engagement in social interaction and that allows for minds and brains to interact with the world not only through signs and symbols but also through hypothesized dimensions of preconscious, unintentional, non-representational, pre-personal modes of neural resonance and mirroring mechanisms in social behavior (Matusall, 2013; Przyrembel et al., 2012; Young, 2012).

Analyzing this recent research, various neuroscientists, STS scholars and philosophers have voiced concerns regarding social neuroscience. There are doubts, for example, about the ecological validity of the way neuroscience operationalizes the social world in a laboratory context (Leys, 2010; Rose & Abi-Rached, 2013a). In turn, there are those who particularly point to the potential interface between conceptualizing the 'social brain' and biopolitical regimes of optimization (Matusall et al., 2011; Rose & Abi-Rached, 2013b; Väliaho, 2014). If the individual's brain is now understood as a potentially 'social brain', then individual brains need to be improved to work towards sociality, to integrate the subject in social life, to become empathic, committed to others, to become 'sociable'. The social brain, on this account, connects to an omnipresent optimization paradigm, as Jan Slaby and Shaun Gallagher argue, 'the particular construal of *self* currently championed by social neuroscience – with a focus on automatic mirroring, low-level empathy and mind-reading – neatly corresponds with the ideal skill profile of today's corporate employee' (2015, p. 45). What becomes clear from these voices, as part of a more general backlash against certain strands of neuroscientific research, is that brain science in general, and social brain science in particular, is a field under particular scrutiny for its imagined social and political consequences.

Art-Science Experiments in Times of Neuro-mania

If neuroscience, and social neuroscience in particular, is indeed a contentious terrain, what could be the position of art-science experiments in relation to this scientific discipline? Some artworks may be seen as fortifying the aforementioned materialist tendencies ('apparent materialization') of brain imaging. For example, when artists use EEG technology to create 'brain paintings' as direct materializations of a cognitive, artistic process, this materialist premise is not questioned but stands at the basis of the artistic work (Hoesle, 2014). On the other hand, when artists put on display such apparent materializations of creative thoughts and dreams, this gesture may also be interpreted differently. For example, in 1963, the renowned American artist Robert Morris created the work *Self-Portrait EEG* by recording his brainwaves while thinking about himself, cutting the paper record of the registered traces when it reached the length of his body. Morris' self-portrait can be viewed as a critical-poetic observation about the impossibility to capture lived experience within the lines of the electroencephalograph, exposing the hubris of this scientific, quantifying enterprise (Weiss, 2014).

Forty years after Morris' EEG artwork however, in today's cultural context of neuro-centrism, we may be more suspicious of public displays of beautiful visualizations of the brain. As the curators of the 2012 exhibition *Brain: The Mind as Matter* (Wellcome Collection, London) point out, today, displays of brains will instantly become part of a 'political economy of brains', whether 'the curator goes with the ideological grain or against it' (Kwint & Wingate, 2013, p. 196). Inevitably, some say, the public presence of the iconic, beautifully exhibited brain feeds into what Nikolas Rose and Joelle Abi-Rached have called 'pedagogies of "brain awareness" ': that is, the multifarious ways in which (often government-sponsored) media events, exhibitions, workshops and public installations may prime laymen audiences to become 'ready to talk' about the brain and make them inclined to understand everyday behavior from a neurological perspective, perhaps even encourage us to work on the brain for self-improvement (Ortega & Vidal, 2011; Pickersgill et al., 2011, p. 357; Rose & Abi-Rached, 2013a, p. 22). Understood as such, the presentation of the brain in public spaces, immediately leads to the aforementioned logics of accountability (1) and logics of innovation (2), aligning an audience with technoscientific developments and biopolitical regimes.

When attending to the question of art's position in times of neuro-mania, it is indeed important not to overlook this element of technoscientific participation, especially because EEG art-science installations often make use of consumer-grade neurotech ware. The data-abstraction and -interpretation methods of these commercial headsets are contested; EEG neuro-technologies are characterized by a speculative nature, in which the full range of technological application is projected to the future.[2] Art-science experiments can contribute to the promissory aspect of these technologies. When artworks result in beautiful images of cyborg-looking people, these can serve as direct promotion for a tech-company's public image. Or, more indirectly, aesthetic displays of neuro-data may also lay the groundwork for our acceptation and readiness to participate in quantifying practices, to use neuro-tools to monitor and manage our style of living (Brenninkmeijer & Zwart, 2017). As such, EEG art-science experiments always participate in a logics of accountability and innovation, in the sense that they make us engage with the desires of translational neuroscience and make us take part in a technology-supported optimization paradigm that asks us to be attentive, relaxed or attuned (Schmitz, 2016).

Yet, while attention to art's relation to technoscientific innovation is important, in this chapter I want to resist a narrow reading of EEG artworks as mere expressions of and drivers of neuro-hype, neuro-centrism or neuro-governmentality, taking seriously the continuously interlocking positions of affect, accountability, innovation and ontology. The history of EEG art-science experiments, starting in the 1960s, demonstrates the way artists have long been able to subvert a straightforward logics of innovation (2), designing circuits between devices and subjects that defy the purpose of reaching 'full potential' or successful behavioral modulation. Artist David Rosenboom (1976) for example, created works in which relations between machines and participants are always messy: circuits are continuously disturbed and irritated, never reaching a fixed state, but precisely drawing attention to the indeterminacy and contingency of this entangled event (Lysen 2019). Over the past two decades, a large community of DIY EEG tech-developers, hackers and artists has continued to experiment with these messy circuits (often directly paying homage to older works of artists such as David Rosenboom and Alvin Lucier), challenging and exposing the black-boxed assumptions in (patented) EEG software (Gastfall et al., n.d.; see Väljamäe et al., 2017; Wexler, 2017).

Such (art) historical observations and recent neuro-art-hacking developments can interface with recent STS scholarship that has called for closer attention to particular situated encounters with technological mediations of the human body and brain and for a detailed analysis of the way 'brainfacts' are produced in practice. What is important, as sociologist Des Fitzgerald (2014) points out, is to closely examine the translational imperative of neuro-imaging technologies – not to immediately accuse such practices of factoring out complexity and to dismiss technical visualizations, such as fMRI scans of the brain, as image-powered reductionism – but to see how puzzling observations and ambiguities are incorporated in the process of knowledge production.

Public art-science experiments can play an important part in this approach to neuroscientific research – close attention to the production of brainfacts in practice – because, during art-science experiments, unruly measurements, contingent elements and unusual behaviors are on display as part of the process of scientific, experimental work. This urgent direction towards new ways of analyzing and doing neuroscientific experimental practice, akin to the logics of ontology (3), is met with recent attempts to form interdisciplinary research teams in neuroscience. Scholars Felicity Callard and Des Fitzgerald have described these current, heterogeneous efforts in various European research projects to bring together scholars from the humanities, social sciences and neurosciences to design what they call 'experimental entanglements' that 'reshape and reimagine the conceptual and empirical contours' of cognitive neuroscientific experiments (Callard & Fitzgerald, 2016, p. 205). Experimental entanglements, as they explain, aim to 'momentarily think something exterior to both the conventions of experimental practice, *and* the taken-for-granted dynamics of epistemic power that underwrite its conduct' (Fitzgerald & Callard, 2015, p. 18). Such moments of subversion and novelty can emerge, for example, by changing the usual methods and conventions of the experiment, by swapping the roles of researchers or altering the way data are interpreted or publicized. In the case of cognitive neuroscientific experiments, such subversions of experimental practice may perhaps 'intensify' and 'magnify' the untidy and messy ways of experimental (neuroscientific) knowledge-production and ultimately 'expose new biosocial stories': that is to become aware, within the space of the experiment, of the entangled nature of categories such as the 'social', 'cultural' or 'biological' and the entanglement of disciplinary practices that were previously viewed as divided and separate (ibid.).

So far, artistic collaborations have rarely played a part in this recent scholarly move towards 'experimental entanglements', yet one can start to see how this new direction paves the way for such art-science involvement. To pay attention to the possibilities of 'experimental entanglements' is to take seriously what Roepstorff and Frith have called the 'aesthetics of research practice' in doing neuroscientific experiments, to be attentive to the configuring elements that are generative of 'new ways of writing, new ways of being in the field, or novel forms of intervention' (Roepstorff and Frith, 2012, p. 103; cited in Fitzgerald and Callard, 2015, p. 22). Analyzing art-science experiments, in this chapter, as an extension of this movement of 'experimental entanglements', we may start to ask what special powers art-science entanglement may have in magnifying and intensifying the practice of neuroscientific experimentation and reconfiguring the phenomena under investigation. A closer view of *EEG KISS* and *Mutual Wave Machine* helps to understand how such reimaginings and reconfigurations can take place.

EEG KISS: Playful Hubris and Radical Relationality

For the art-science experiment *EEG KISS*, artists Maat and Lancel (2014) invite two participants at a time to wear an EEG headset and kiss each other. Audience members may watch the kissers during their act and are able to witness changing inscriptions of EEG activity on an adjacent computer screen or large projection. Over the past years, hundreds of participants have thus created 'digital portraits' of their kisses during installments at art galleries, art-science festivals, tech-industry events and scientific conferences. The work invites lovers (and strangers) to put their kisses on double display: to expose an intimate act to a public audience and to show a data-visualization of this act in a public environment (Plate 16). As such, *EEG KISS* creates a contradicting and potentially subversive environment that raises a number of important issues, such as epistemological questions relating to the measurability of complex behavioral phenomena (is it possible to quantify a kiss or a feeling of intimacy?), ethical questions regarding the relation between public data and private emotions (what if emotion-data would become public property?) and phenomenological issues (how do data-visualizations effect our experience of intimacy? Does the nature of intimacy itself change through new technologies?) The artwork's ambition to quantify the kiss – one of the most intimate, emotional and complex behavioral phenomena – can be understood as a playful gesture, one that makes tangible the potential hubris of neuro-scientific research. The artwork seems to point to the direction of inevitable failure of this quantifying enterprise and exposes our longstanding desire to record and visualize cognition, emotion and memory within scientific registers.

As a public experiment, *EEG KISS* can be understood as enacting the logics of ontology in a very literal sense when the artists describe the work as a 'social lab' in which audience members become 'co-researchers' in the art-science experiment (Maat & Lancel, 2014). Juxtaposing their work to a scientific lab, this 'social', art-science lab thus exposes the conditions of researching intimacy, the way different lab configurations (presence of an audience or the creation of an atmosphere) may radically change measurements and interpretations. Such subversions prompt a number of questions: what is a 'natural' kissing situation, how can we redefine intimacy? How does a view of my data influence my (feeling of) this kiss? *EEG KISS*, on this account, aims to help participants and viewers study and understand the social relations, aesthetic choices, performative scripts, assumptions, errors and coincidences that are part of all scientific practices, but that become more apparent by staging an experiment in an artistic context. *EEG KISS*, as an art-science experiment, reconceptualizes the object of research – 'intimacy' – by reconfiguring what intimacy is in the interplay between public and private experiences, feelings and displays. As such, this art-science project may indeed be viewed as generating 'something new within scientific practice itself, challenging the boundaries of disciplinary authority': *EEG KISS* reimagines the way intimacy could itself be enacted within research about the social brain (Born & Barry, 2010, p. 114).

A second outcome of this specific art-science experiment into the social brain is that *EEG KISS*, through its performative design, emphasizes and makes us aware of the continuous interrelation between the visualized technological measurements, the experimental environment (with an audience present), individual experiences of participants and the dyadic behavior of the two kissers. Participants are influenced by sensations of bodily presence of spectators, by the sounds and images of fluctuating neural activity, which in turn, impacts brain activity and alters these scientific inscriptions. Viewed as such, this art-science experiment may be interpreted as staging the on-going feedback loop between brains, mediated brains, bodies and world. A certain relationality – a feedback-natured system – is made sensible to participants and put on spectacular display for spectators who can view kissing bodies, computer screens and brain inscriptions in continuous exchange. Accordingly, artistic experiments with EEG technologies can potentially create situations in which brain activity is measured and revealed as enmeshed in material environments, fleshy bodies and intersubjective encounters, to offer a vision of what philosopher Protevi (2010, p. 170) has described as the 'radical relationality' of the 'brain–body–world system'. 'Radical relationality means that being human is composed of relations; we do not "have" relations, but we are relations all the way down' (ibid., p. 174). Such a view steers away from neurocentrism, and adheres to a view of the brain as one among other 'mediating' or 'relational' organs (Fuchs, 2011, p. 198; De Jaegher, 2018).

While neuro-philosophers reflect on neuroscientific practice to propose such intersubjective theories of social cognition, this art-science experiment adds to this movement differently by offering an experience of the continuous interplay between humans, machines and materials and putting this interplay on spectacular, performative display. On this view, some art-science works have a special reflective-affective capacity, enabling new thoughts and realizations of our embodied, embrained, enworlded being and of the entangled nature of neuroscientific experiments. Philosopher of science Andrew Pickering has usefully described the way artworks can thus function as 'philosophical objects', pointing 'towards rather different conceptions of being in the world' (2008, p. 1). Seen as such, *EEG KISS* engenders – with humor, theatricality and saliva – a site of performance, a space of sensuous entanglement that foregrounds the logics of ontology and affect.

Mutual Wave Machine: 'Dark Matter' of Brain Broadcasting

Similar to *EEG KISS*, the art-science experiment *Mutual Wave Machine* can be understood as enacting the logics of ontology and affect: the work reconfigures what it means to be or to feel in 'synchrony' with another person. The artwork demands of its participants to 'try to be on the same wavelength' when they enter the installation, an egg-shaped dome in which two subjects wearing EEG caps can sit opposite each other in mutual concentration. Created by cognitive scientist Susanne Dikker and artist Matthias Oostrik, *Mutual Wave Machine* is presented as an art installation that offers an eye-catching participatory sensory experience with dramatic lighting, illuminated visualizations and ambient sounds. Installed in various artistic and scientific contexts, the measured EEG data of this art-science experiment also feed into an explorative scientific study, part of an emerging cognitive science research program in 'second person neuroscience', a field that forms hypotheses about the neural mechanisms that are involved in social situations and person-to-person interaction. New technological abilities to produce brain images of active brains of two people at the same time (a technique called 'hyperscanning') and new paradigms in EEG-technology have allowed for a new examination of what is called 'brain-to-brain coupling' (states of brainwave synchronization between participants).[3]

Mutual Wave Machine asks 'what can the brain tell us about what it means to be "on the same wave length" with another person?', hypothesizing that 'your brainwaves are more in sync with those of another person when you are feeling connected to that person' (Dikker et al., 2014).[4] While partaking in the experiment, participants and spectators are able to see a real-time visual feedback on the achieved levels of 'brainwave synchrony': the more synchrony is measured, the more each side of the dome will show a video projection of the participants' faces. This visual articulation of synchrony is hypothetical and speculative, yet it has strong reality effects. Even though it is a visualization of emergent data interpretation, the feedback imagery achieves a level of scientific veracity, a playful 'apparent materialization' of synchrony.

Beyond its speculative and spectacular visual feedback, it is through the logics of affect that the installation has remarkable effects. The dome offers a space for experimenting in subtle forms of contact: participants are invited to experiment with – and to become more aware of – the very delicate attunements that take place in a person-to-person situation. It is the mutual gazing, speaking, listening and gesturing that causes neural patterns of attunement, patterns that can be formed – so the second person neuroscience paradigm proposes – through subtle and preconscious dynamics of perception and action: that is, the 'continuous moment-to-moment mutual adaptation of our own actions and the actions of the other' (Konvalinka & Roepstorff, 2012, p. 2). *Mutual Wave Machine* enacts the logics of affect by offering new sensibilities: that is, novel possibilities for the body's capacity to act in relation to others.

As an art-science experiment, *Mutual Wave Machine* creates an environment in which to explore and become newly aware of this everyday phenomenon of interpersonal synchronicity. At the same time, its aesthetic design and media-technical environment also imbue our understanding of synchronicity with a certain ambiguity. This ambiguity is telling for this emerging scientific research program: when first encountering the term 'brain-to-brain coupling' in an academic paper, it may conjure visions of science-fictional, ethereal contact between brains, a media analogy of brains that communicate over a distance through some form of mental radio broadcasting. Watching the gawking spectators around *Mutual Wave Machine*, it is reasonable to say that this speculative and fantastic imaginary of 'psychic' brainwave communication is strengthened by the installation and successfully attracts visitors. Some intuitive interpretation of thought transmissions is fortified, for example, by digital animations (accompanying the installation) of illuminated flashes traversing a dark space between two glowing brains (Figure 13.2). With these elusive hints, the installation readily evokes longstanding historical imaginaries of mindreading and telepathic transmission.

Reacting against such fantastic interpretations, scientists have urged understanding brain-to-brain coupling not as a type of psychic connection via invisible brainwaves but as shared brain-activity patterns between participants, emerging from shared sensations of and reactions to signals in the environment, most importantly, from the mutual moment-to-moment sensing of and adapting to one another. As cognitive neuroscientist Guillaume Dumas explains, 'there is nothing magical here', it is through the action-perception flow that people form a coherent system: 'each exchange is an opportunity to overcome our individuality', and it is in this non-magical yet enthralling way that 'we are more connected than we think' (Dumas, 2011; Dumas & Halard, 2012).

Nevertheless, even though scientists ward off 'magical' interpretations of second-person neuroscience, something of the media imaginary of a brainwave broadcasting space persists. Both artists and scientists use the media-technical analogy of a broadcasting situation to describe the affective state of the participants ('staying tuned', 'ticking together', 'being on the same wavelength'), and they use the same analogies to portray the neural dimension of brainwave attunement.[5] Moreover, in scientific publications, wave synchrony between similar areas of two brains is visually represented by drawing lines from one brain to another brain, thus constructing a graphic 'bridge' that denotes a measured level of synchrony between brain areas of different subjects (Plate 17).[6] What happens, through these graphic and discursive configurations, is the construction of an intuitive association between the person-to-person process of tuning and states of neural activity. In this way, the hypothesized connection between sensed synchrony and measures of brainwave synchrony is rhetorically fortified. This material-discursive configuration of brain-to-brain coupling is not a performative liberty on the part of art-science, but an integral part of this scientific research program.

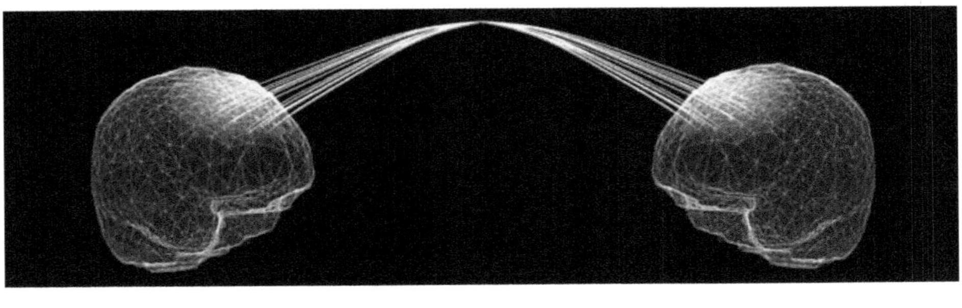

Figure 13.2 Still from visuals used in *Measuring the Magic of Mutual Gaze*. Suzanne Dikker & Matthias Oostrik.

When art-science experiments or scientific papers imagine interpersonal contact as a type of broadcasting situation, this vision elides the explanatory gap between the level of social-affective experience ('feeling in sync') and the measured similarities in neural oscillations (brainwave synchrony). Hence, the work this analogy performs goes to the heart of the enactivist critique of second-person neuroscience, namely that it lacks a theory to connect the subpersonal, neuronal level to the personal level of meaning and experience (Sameen et al., 2013). As such, the broadcasting analogy may steer towards a representational cognitive science paradigm, critiqued by enactive theorists as perpetuating the idea of detached, information-processing minds. Yet, here I propose that *Mutual Wave Machine*, as an art-science experiment, again intervenes in this field through logics of affect by intensifying and magnifying this scientific ambiguity in sensuous ways.

By casting this art-science experiment in a dramatic atmosphere of mysterious, uncanny forces, the installation subverts the broadcasting analogy in the context of a neuroscientific experiment, giving shape to a long-existing imaginary that points towards the conceptual fragility of boundaries between people, spaces and machines. Indeed, historically, it is exactly the idea of a vast, all-encompassing (radio) broadcasting space that has caused desires and anxieties of transgressing everyday borders.[7] As a type of science fictional device, the *Mutual Wave Machine* asks us to think about these persistent imaginaries of para-scientific communication and about the way science is haunted by such lingering supernatural fantasies. The evocative, aesthetic design of this machine generates a thought-provoking disjunction between an awe-inspiring mind-reading machine and the subtleties of genuine, everyday person-to-person contact. It is this subversive disjuncture that finally makes us aware of the uncertainties in the domain of social neuroscience, as a recent team of neuroscientists, psychologists and philosophers put it, 'true social interaction remains the "dark matter" of social neuroscience' (Przyrembel et al., 2012). Ultimately, it is the search for new experimental (more ecological, more sensitive) test environments that can take this research further, and it is precisely in this search that art-science experiments intervene. *Mutual Wave Machine*, as a science-fictional medium that invokes dark matter, asks us to reconsider how we should envision and test person-to-person interaction in social neuroscience.

Conclusion: Intersecting Logics and New Neuro-enchantment

In this chapter, two art-science experiments in the field of social neuroscience reveal different ways in which hybrid intersections between the field of EEG research and artistic experiments generate reflections on and reconfigurations of the 'social brain' under investigation, showing new perspectives and new methods of research in social neuroscience. *EEG KISS* questions and reshapes the concept of 'intimacy' through the public and performative setting of a scientific experiment. Moreover, this art-science experiment functions as a special, onto-affective performative space to put on display the relationality of behavior, environment and media-technical inscriptions. In turn, the art-science experiment *Mutual Wave Machine* complicates and opens up what we may mean by person-to-person synchrony and 'being on the same wavelength'. Throughout this art-science event, viewers and participants are prompted to watch and experience interpersonal contact differently, with new scrutiny for movement, pace and intensity. Additionally, the artful design of this science-fictional 'machine' asks for a closer look at the media-technical analogy of personal contact as a type of broadcasting situation.

On another concluding note, it is important to stress that part of the intensifying, magnifying and subverting powers of these art-science experiments is not always immediately sensible to first-time participants. Moments of novelty and exposed entanglements within the art-science experiment often emerge over a longer period of time, with changing configurations of software, adaptions to the performative set-up and collected interview data. When participants start wearing an EEG headset, they too, can become experimenters but only for a brief moment in time and as part of a longer process of experimental investigation. The two case studies in this chapter offer examples of art-science installations in which the participating members of the public alternate positions as spectators, as experimenters and as objects of investigation. Participants may not always be aware of these positions. One can be entangled within a subversive reconfiguration of intimacy for neuroscience and at the same time overwhelmed by the beautiful sounds of an assumed auditory portrait of a romantic kiss. Different temporalities of experimentation and different logics of interdisciplinarity can be simultaneously present within art-science experiments.

Importantly, the heterogeneity of audience positions makes an important amendment to a common invocation of an all too eager and credulous 'general public' for neuroscience. Various scholars claim that laymen may be easily persuaded, with the help of neuro-images, to overestimate the present state of knowledge gained by neuroscience and to cast a neuro-centrist perspective on behavior.[8] 'Neuro-enchantment' is the lovely but pejorative term given to this 'sub-judicious fascination with brain science', from which the public should be safeguarded, so it is claimed, by education in critical thinking (Ali et al., 2014, p. 1). Veering away from this pre-defined naïve general public, the two examples of art-science experiments in this chapter do not interpellate the participating members as mere gullible science spectators or as 'walking brains' for data-acquisition but as part of an entangled experiment in which subject and object positions may continuously change. This does not mean that dimensions of neuro-enchantment are not present within art-science experiments, but rather, that enchantment itself is an integral part of their intensifying, magnifying and subverting powers.

To conclude, the two cases studies of art-science experiments discussed show that dimensions of accountability, innovation, ontology and affect always intersect. In the case of *EEG KISS* for example, sponsored EEG software is both improved and advertised through the work, while at the same time, underlying assumptions on what counts as valid measurements of the 'intimate brain' within neuroscientific research are brought to the fore. Hence, art-science experiments offer a space in which it is impossible to separate these 'logics of interdisciplinarity': dimensions of ethical commentary, social engagement, stimulating marketability, affective affordances and reconceptualizing experimental definitions are continuously entangled.

Through the effects of these logics, the art-science experiments here discussed make a critical intervention into social neuroscience research. Such a critique is not pre-conceived – hence, not a prefigured position of art as critical commentator – but is rather, as science and technology scholars have remarked, an 'emergent' form of critique, not art primarily aimed at commenting on pre-defined aspects of neuroscience but an open-ended type of questioning that arises from new configurations (Calvert & Schyfter, 2016). Kissing and staring generate new relations in art-science experiments, challenging the boundaries – and comfort zones – of disciplinary authority.

Acknowledgements

I thank Suzanne Dikker, Karen Lancel and Guillaume Dumas for their helpful comments on this article.

Notes

1 For a good overview of different types of objections against neuroscience, see Meloni, 2011. On 'neuro-governmentality', see particularly Rose & Abi-Rached, 2014.
2 There is no consensus about the accuracy of the standardized bandwidths for measuring brain activity or about the definitions of the measured states of brain activity (Chu, 2015). About the promissory nature of neurotechnologies, see Gardner & Wray, 2013; Martin, 2015.
3 Research using hyperscanning of two brains is still in an explorative phase, little is known about the significance and nature of what is variably called 'brain-to-brain coupling', 'inter-brain phase synchronization', 'neural synchronization', 'neural coupling', 'brain-to-brain synchrony' or 'real-time neural dynamics' (Dikker et al., 2014; Dumas et al., 2010; Sänger et al., 2011).
4 In the experiment, two subjects are asked to try to 'synchronize' their brain activity with the activity of the person facing them in a time period of about 10 minutes. The intake-form suggests the subjects might try to do this, for example, by thinking about similar things.
5 Cognitive neuroscientist Uri Hasson, for example, describes brain-to-brain coupling as 'analogous to a wireless communication system in which two brains are coupled via the transmission of a physical signal' (Hasson et al., 2012, p. 115).
6 See, for example, images in Dumas et al., 2010; Lindenberger et al., 2009, p. 22.
7 Media scholar Jeffrey Sconce has described a historical genealogy of a 'logic of transmutable flow' (2000, p. 8), enabled by public imaginations of media such as radio and television, which enabled the metaphorical assemblage of electricity, information and consciousness, this 'logic of flow' underpins discourses on the boundary transgressing qualities of these media.
8 Some of the given explanations of this attraction to neuro-centrism are 'people's natural affinity for reductionistic explanations of cognitive phenomena' or a 'particular willingness, a specific kind of satisfaction gained in construing oneself as powerless, a-rational, subject to uncontrollable external or internal forces that ultimately determine one's fate' (McCabe & Castel, 2008, p. 344; Slaby & Gallagher, 2015, p. 46).

References

Ali, S. S., Lifshitz, M., & Raz, A. (2014). Empirical Neuroenchantment: From Reading Minds to Thinking Critically. *Frontiers in Human Neuroscience, 8*, Article 357.

Anderson, B. (2007). Hope for Nanotechnology: Anticipatory Knowledge and the Governance of Affect. *Area, 39*(2), 156–165.

Beaulieu, A. (2002). Images Are Not the (Only) Truth: Brain Mapping, Visual Knowledge, and Iconoclasm. *Science, Technology and Human Values, 27*(1), 53–86.

Benschop, R., Peters, P., & Lemmens, B. (2014). Artistic Researching: Expositions as Matters of Concern. In M. Schwab & H. Borgdorff (Eds.), *The Exposition of Artistic Research: Publishing Art in Academia* (pp. 39–51). Amsterdam: Amsterdam University Press.

Borgdorff, H. (2010). The Production of Knowledge in Artistic Research. In M. Biggs & H. Karlsson (Eds.), *The Routledge Companion to Research in the Arts* (pp. 44–63). London and New York: Routledge.

Born, G., & Barry, A. (2010). Art-Science: From Public Understanding to Public Experiment. *Journal of Cultural Economy, 3*(1), 103–119.

Brenninkmeijer, J., & Zwart, H. (2017). From 'Hard' Neuro-Tools to 'Soft' Neuro-Toys? Refocussing the Neuro-Enhancement Debate. *Neuroethics, 10(3)*, 337–348.

Callard, F., & Fitzgerald, D. (2015). *Rethinking Interdisciplinarity across the Social Sciences and Neurosciences.* New York: Palgrave Macmillan.

Callard, F., & Fitzgerald, D. (2016). Entangling the Medical Humanities. In A. Whitehead, A. Woods, S. J. Atkinson, J. Macnaughton & J. Richards (Eds.), *The Edinburgh Companion to the Critical Medical Humanities* (pp. 35–49). Edinburgh: Edinburgh University Press.

Calvert, J., & Schyfter, P. (2016). What Can Science and Technology Studies Learn from Art and Design? Reflections on 'Synthetic Aesthetics'. *Social Studies of Science, 47*(2), 195–215.

Choudhury, S., & Slaby, J. (Eds.). (2011). *Critical Neuroscience: A Handbook of the Social and Cultural Contexts of Neuroscience.* Oxford: Wiley-Blackwell.

Chu, N. N. Y. (2015). Brain-Computer Interface Technology and Development: The Emergence of Imprecise Brainwave Headsets in the Commercial World. *IEEE Consumer Electronics Magazine, 4*(3), 34–41.

De Jaegher, H. (2018). The Intersubjective Turn. In A. Newen, L. De Bruin & S. Gallager (Eds.), *The Oxford Handbook of 4E Cognition* (pp. 453–468). Oxford: Oxford University Press.

Dikker, S., Kirnos, K., & Oristaglio, S. (2014). Neuroscience Exclusives. Digital MAI Exclusives Package. Retrieved from https://www.kickstarter.com/projects/maihudson/marina-abramovic-institute-the-founders/posts/764797.

Dikker, S., Silbert, L. J., Hasson, U., & Zevin, J. D. (2014). On the Same Wavelength: Predictable Language Enhances Speaker–Listener Brain-to-Brain Synchrony in Posterior Superior Temporal Gyrus. *Journal of Neuroscience, 34*(18), 6267–6272.

Dumas, G. (2011). Towards a Two-Body Neuroscience. *Communicative and Integrative Biology, 4*(3), 349–352.

Dumas, G., & Halard, L. (2012). Phi (subtitled). Retrieved from https://www.youtube.com/watch?v=PPbH6CMw2bU.

Dumas, G., Nadel, J., Soussignan, R., Martinerie, J., & Garnero, L. (2010). Inter-Brain Synchronization during Social Interaction. *PLoS ONE, 5*(8), e12166.

Fitzgerald, D. (2014). The Trouble with Brain Imaging: Hope, Uncertainty and Ambivalence in the Neuroscience of Autism. *BioSocieties, 9*(3), 241–261.

Fitzgerald, D., & Callard, F. (2015). Social Science and Neuroscience beyond Interdisciplinarity: Experimental Entanglements. *Theory, Culture and Society, 32*(1), 3–32.

Fuchs, T. (2011). The Brain, a Mediating Organ. *Journal of Consciousness Studies, 18*(7–8), 196–221.

Gardner, P., & Wray, B. (2013). From Lab to Living Room: Transhumanist Imaginaries of Consumer Brain Wave Monitors. *ADA: Journal of Gender, New Media and Technology, 3.* Retrieved from https://adanewmedia.org/2013/11/issue3-gardnerwray.

Gastfall, U., Gustin, P., & Parésys, G. (n.d.). Journal du projet EEG: In-Between. Retrieved from http://gerard.paresys.free.fr/Projets/ProjetEEG.html.

Gieryn, T. (1983). Boundary-Work and the Demarcation of Science from Non-Science: Strains and Interests in Professional Ideologies of Scientists. *American Sociological Review, 48*(6), 781–795.

Halpern, M. K. (2012). Across the Great Divide: Boundaries and Boundary Objects in Art and Science. *Public Understanding of Science, 21*(8), 922–937.

Hasson, U., Ghazanfar, A. A., Galantucci, B., Garrod, S., & Keysers, C. (2012). Brain-to-Brain Coupling: A Mechanism for Creating and Sharing a Social World. *Trends in Cognitive Sciences, 16*(2), 114–121.

Hoesle, A. (2014). Between Neuro-potentials and Aesthetic Perception: Pingo Ergo Sum. In G. Grübler & E. Hildt (Eds.), *Brain-Computer-Interfaces in Their Ethical, Social and Cultural Contexts* (pp. 99–108). Dordrecht: Springer.

Jones, C. (2012). Experience. In A. v. d. Heiden & N. Zschocke (Eds.), *Autorität des Wissens Kunst- und Wissenschaftsgeschichte im Dialog* (pp. 167–191). Zürich: Diaphanes.

Konvalinka, I., & Roepstorff, A. (2012). The Two-Brain Approach: How Can Mutually Interacting Brains Teach Us Something about Social Interaction? *Frontiers in Human Neuroscience, 6*, Article 215.

Kwint, M., & Wingate, R. (2013). Curating the Brain. *Interdisciplinary Science Reviews, 38*(3), 195–199.

Lapworth, A. (2016). Theorizing Bioart Encounters after Gilbert Simondon. *Theory, Culture and Society, 33*(3), 123–150.

Leys, R. (2010). How Did Fear Become a Scientific Object and What Kind of Object is It? *Representations, 110*(1), 66–104.

Lindenberger, U., Li, S.-C., Gruber, W., & Müller, V. (2009). Brains Swinging in Concert: Cortical Phase Synchronization while Playing Guitar. *BMC Neuroscience, 10*(1), 22.

Littlefield, M. (2017). 'A Mind Plague on Both Your Houses': Imagining the Impact of the Neuro-Turn on the Neurosciences. In J. Leefmann & E. Hildt (Eds.), *The Human Sciences after the Decade of the Brain* (pp. 198–213). London: Academic Press.

Lysen, F. (2019). The Interface is the (Art) Work: EEG-Feedback, Circuited Selves and the Rise of Real-Time Brainmedia (1964–1977). In A. Nijholt (Ed.), *Brain Art: Brain–Computer Interfaces for Artistic Expression*. Human-Computer Interaction Series. Cham: Springer.

Maat, H., & Lancel, K. (2014). *EEG KISS*. Retrieved from http://www.lancelmaat.nl/work/e.e.g-kiss.

McCabe, D. P., & Castel, A. D. (2008). Seeing is Believing: The Effect of Brain Images on Judgments of Scientific Reasoning. *Cognition, 107*(1), 343–352.

Martin, P. (2015). Commercialising Neurofutures: Promissory Economies, Value Creation and the Making of a New Industry. *BioSocieties, 10*(4), 422–443.

Massumi, B. (2015). *Politics of Affect*. Cambridge: Polity Press.

Matusall, S. (2013). Social Behavior in the "Age of Empathy"? A Social Scientist's Perspective on Current Trends in the Behavioral Sciences. *Frontiers in Human Neuroscience, 7*, Article 236.

Matusall, S., Kaufmann, I. M., & Christen, M. (2011). The Emergence of Social Neuroscience. In J. Decety & J. T. Cacioppo (Eds.), *The Oxford Handbook of Social Neuroscience* (pp. 9–27). Oxford: Oxford University Press.

Meloni, M. (2011). Philosophical Implications of Neuroscience: The Space for a Critique. *Subjectivity, 4*(3), 298–322.

Meynell, L. (2012). The Politics of Pictured Reality: Locating the Object from Nowhere in fMRI. In R. Bluhm, A. J. Jacobson & H. L. Maibom (Eds.), *Neurofeminism: Issues at the Intersection of Feminist Theory and Cognitive Science* (pp. 11–29). New York: Palgrave Macmillan.

Michael, M. (2012). "What Are We Busy Doing?" Engaging the Idiot. *Science, Technology, and Human Values, 37*(5), 528–554.

Ortega, F., & Vidal, F. (2011). Toward a Genealogy of Neuroascesis. In F. Ortega & F. Vidal (Eds.), *Neurocultures: Glimpses into an Expanding Universe* (pp. 27–44). Frankfurt am Main and New York: Peter Lang.

Park, L. (2017). Eunoia. Retrieved from http://www.thelisapark.com/eunoia.

Pickering, A. (2008). New Ontologies. In A. Pickering & K. Guzik (Eds.), *The Mangle in Practice: Science, Society, and Becoming* (pp. 1–14). Durham, NC, and London: Duke University Press.

Pickersgill, M., Cunningham-Burley, S., & Martin, P. (2011). Constituting Neurologic Subjects: Neuroscience, Subjectivity and the Mundane Significance of the Brain. *Subjectivity, 4*(3), 346–365.

Protevi, J. (2010). Deleuze and Wexler: Thinking Brain, Body and Affect in Social Context. In D. Hauptmann & W. Neidich (Eds.), *Cognitive Architecture: From Bio-politics to Noo-politics: Architecture and Mind in the Age of Communication and Information* (pp. 168–183). Rotterdam: 010 Publishers.

Przyrembel, M., Smallwood, J., Pauen, M., & Singer, T. (2012). Illuminating the Dark Matter of Social Neuroscience: Considering the Problem of Social Interaction from Philosophical, Psychological, and Neuroscientific Perspectives. *Frontiers in Human Neuroscience, 6*, Article 190.

Roepstorff, A., & Frith, C. (2012). Neuroanthropology or Simply Anthropology? Going Experimental as Method, as Object of Study, and as Research Aesthetic. *Anthropological Theory, 12*(1), 101–111.

Rose, N., & Abi-Rached, J. M. (2013a). *Neuro: The New Brain Sciences and the Management of the Mind*. Princeton, NJ: Princeton University Press.

Rose, N., & Abi-Rached, J. M. (2013b). The Social Brain. In *Neuro: The New Brain Sciences and the Management of the Mind* (pp. 141–163). Princeton, NJ: Princeton University Press.

Rose, N., & Abi-Rached, J. (2014). Governing through the Brain: Neuropolitics, Neuroscience and Subjectivity. *Cambridge Journal of Anthropology, 32*(1), 3–23.

Rosenboom, D. (1976). *Biofeedback and the Arts: Results of Early Experiments*. Vancouver: ARC Publications.

Roskies, A. L. (2010). Neuroimaging and Inferential Distance: The Perils of Pictures. In S. J. Hanson & M. Bunzl (Eds.), *Foundational Issues in Human Brain Mapping* (pp. 195–216). Cambridge, MA: MIT Press.

Salter, C., Burri, R. V., & Dumit, J. (2016). Art, Design, Performance. In U. Felt, R. Fouché, C. A. Miller & L. Smith-Doerr (Eds.), *The Handbook of Science and Technology Studies* (pp. 139–168). Cambridge, MA: MIT Press.

Sameen, N., Thompson, J., & Carpendale, J. (2013). Further Steps toward a Second-Person Neuroscience. *Behavioral and Brain Sciences, 36*(4), 437.

Sänger, J., Lindenberger, U., & Müller, V. (2011). Interactive Brains, Social Minds. *Communicative and Integrative Biology, 4*(6), 655–663.

Schmitz, S. (2016). The Communicative Phenomenon of Brain–Computer Interfaces. In V. Pitts-Taylor (Ed.), *Mattering: Feminism, Science, and Materialism* (pp. 140–157). New York: NYU Press.

Sconce, J. (2000). *Haunted Media: Electronic Presence from Telegraphy to Television*. Durham, NC: Duke University Press.

Slaby, J., & Gallagher, S. (2015). Critical Neuroscience and Socially Extended Minds. *Theory, Culture and Society, 32*(1), 33–59.

Stelarc (2010). *Spectacle of the Mind* [live performance: audiovisual effects generated from brainwaves using purpose-designed EEG video software and Emotiv EPOC neuro-headset]. Melbourne.

Väliaho, P. (2014). *Biopolitical Screens: Image, Power, and the Neoliberal Brain*. Cambridge, MA: MIT Press.

Väljamäe, A., Evers, L., Z. Allison, B., Ongering, J., Riccio, A., Igardi, I. & Lamas, D. (2017). *The BrainHack Project: Exploring Art-BCI Hackathons*. Retreived from https://www.researchgate.net/publication/314504834_The_BrainHack_Project_Exploring_Art_-_BCI_Hackathons.

Weiss, J. S. (2014). *Robert Morris: Object Sculpture, 1960–1965*. New Haven, CT: Yale University Press.

Wexler, A. (2017). The Social Context of 'Do-It-Yourself' Brain Stimulation: Neurohackers, Biohackers, and Lifehackers. *Frontiers in Human Neuroscience, 11*, Article 224.

Wienroth, M., & Goldschmidt, P. (2017). Facilitating Creative Equality in Art-science: A Methodological Experiment. *Leonardo, 50*(1), 42–46.

Young, A. (2012). The Social Brain and the Myth of Empathy. *Science in Context, 25*(3), 401–424.

Zschocke, N. (2010). Art and Science Research Teams? Some Arguments in Favour of a Culture of Dissent. In J. Scott (Ed.), *Artists-in-Labs Networking in the Margins* (pp. 68–81). Vienna: Springer Verlag.

Zwijnenberg, R. (2012). Two-Headed Zebrafish: Some Personal Observations on the Relevance and Urgency of Collaborations Between Art, Science and Technology. *Moebius, 1*(1), 1–14.

14 Re-enactment as a Research Strategy

From Performance Art to Video Analysis and Back Again

Philippe Sormani

Introduction

Over the last three decades, artistic research, arts-based research, and practice-based inquiry more broadly have received increasing attention in and beyond academia, both epistemologically and institutionally speaking (e.g., Borgdorff, 2012; Sormani et. al., 2018; Wilson & Van Ruiten, 2013). Against this background, the present chapter offers a methodological reflection on re-enactment as a research strategy. For this purpose, the chapter explores the interplay between performance art and video analysis. More specifically, the chapter revisits a particular position in performance art – Andrea Fraser's institutional critique *qua* filmed intervention – in dialogue with practice-based video analysis, a recent development in ethnomethodology (EM) practiced "alongside mainstream STS" (Sormani et al., 2017, p. 123).[1]

Therefore, the chapter examines three media announcements from a developing corpus of video recordings, all of which announce one form or another of "machine intelligence," relating to video gaming, neuromorphic computing, and machine learning, respectively. Just how is machine intelligence announced in each of these cases, if not demonstrated as a credible artifact of sorts? And, secondly, what kind of framings – epistemic, aesthetic, institutional – do the respective announcements imply? These two questions cut across classic STS (Shapin, 1995), media studies (Broth et al., 2014), and critical aesthetics (Rebentisch, 2013). The chapter answers them by analyzing, re-enacting and, on that basis, re-analyzing the selected announcements, thus contributing to a "sociology of demonstrations" (Rosental, 2013; 2017).[2]

Various collaborations in teaching, research, and outreach contexts made possible the chapter's findings, collaborations whose instructive character shall be made explicit in and as the chapter's course. In probing the dramaturgical framing of machine intelligence, the chapter challenges the determinist staging of its technical operation and, for this purpose, spells out the heuristic interest of "trouble" from an ethnomethodological stance (Garfinkel, 1967; 2002). The concluding discussion draws together the key insights from the presented video analysis, both practice-based and frame-explicating, at the tricky intersection of performance art, Fraser's institutional critique, and video analysis, if not contemporary art and current STS more broadly.[3]

Lamenting "Performance Art"? From Performance to Enactment (A. Fraser)

Currently, a paradox seems to haunt the cultural category of "performance art," a paradox that has led internationally renowned performance artist Andrea Fraser to lament the worldly success of the notion of "performance" and the qualifier "performative" in particular (Fraser, 2014). Before turning to Fraser's essay, lament, and alternative in one, it is worth briefly elaborating on the paradox.

Over the last 30 years, the notion of "performance" has seen an expanding usage across various fields of cultural practice and academic interest. Not only has it given rise to new fields of cultural and scholarly engagement, including transdisciplinary fields such as "performance studies" (see Schechner, 2002), but it also has been drawn upon to recast existing fields, ranging from anthropology (Turner, 1998) and sociology (Gergen & Gergen, 2014) to feminism (Butler, 1990), STS (Salter et al., 2017), and feminist STS (Herzig, 2004), as well as dance (Costinas & Janevski, 2017). Paradoxically, this worldly success, no means confined to academia (e.g., Butler, 2015), stands in ironic contrast to lingering self-doubts in the fragmented field of contemporary art, at times culminating in the question "what's wrong with performance art?" (Corner College, 2017).[4]

Pending a comprehensive survey, three recurring lines of self-doubt may be indicated.

First, the *common realization of the lost anti-institutional impetus of performance art*. As RoseLee Goldberg, performance historian (Goldberg, 2011) and founding director of Performa, a festival based in New York City, recently put it: "I am often asked, 'What if performance becomes too institutionalized in the museum context?' But it has been 'institutionalized' for a long time, if we consider the work of some of the major artists" (Goldberg et al., 2014, p. 17). Goldberg's rejoinder, then, expresses perplexity, rather than resignation, at the question still being asked. As she explains with respect to her own curatorial practice, "I soon got over it" (ibid., p. 9) – in contrast, it seems, to many of her interlocutors, real or virtual, including Claire Bishop recently noting a "curious depolitization of performance art" (2017, p. 41).

Second, the *ironic acknowledgment of the lost pertinence of a recurring cliché*, the "cliché about performance art that it involves artists enacting strenuous tasks" (Adamson & Bryan-Wilson, 2016, p. 91). A recent review of Marina Abramović's autobiography (2016) offers a sharp expression of this ironic acknowledgement, as the review deconstructs the artist's tale of self-styled sufferings (many of which have become famous performances) as a ridicule testimony of "masochism and pretension" (Garner, 2016). Only an incidental concession tempers the deconstructive fervor: "a tolerance for a certain amount of pomposity is a prerequisite for keeping up with serious art; otherwise, you're always sitting at the short table and using the plastic cutlery" (ibid.). Deflationary moves thus may not necessarily offer artistic solace either.

Third, the *empty redundancy of the inflationary use of the qualifier "performative"* observed by artists and art historians alike. Whilst Fraser (2014) notes the dissolution of the specificity of the notion of "performance art" as marking a distinctive genre, Dorothea von Hantelmann observes the casual pleonasm in an expression such as "performative art," given that an artwork by definition is always already "performative" (2010, pp. 17–18). Indeed, through its very production and presence, *qua* artwork, it must bring into existence a new object, statement, or reality – in line with John L. Austin's initial characterization of "performatives" as reality-constituting expressions, rather than activity-qualifying adjectives (Austin, 1962).

As Fraser explains, "the term 'performative' [when used as an omnirelevant adjective] has become a kind of camouflage or lure .... Rather than opening up all manner of forms and activities to a reflection on what we do ... that primary meaning of the term is now mostly consigned to anachronistic and academic usage" (Fraser, 2014,

p. 124).[5] Conversely, Fraser's essay suggests to recast performance art (at least as her own work is concerned) in terms of "enactment":

> If we are always enacting, and if these structures [social, psychological, and economic relationships] are always there, performance art – and art generally, as I understand it – aims, first of all, to *occasion a recognition of and reflection on those structures in their enactment*: structures that include not only what artists, performers, or intellectuals do, but what audiences, readers, and other participants in any encounter also do. And this, for me is how "performative", if I used the word, would be defined: as enactment that performs itself and in so doing structures a *recognition of and reflection on the relations produced and reproduced in the activity and, above all, on the investments that orient them.* (Fraser, 2014, p. 127; emphasis added).

As a project rationale, this position statement is both clear and concise, yet potentially ambiguous. Clearly, it seems to be assuming ubiquitous structures, relations, and investments (as being "always there") to a point where any and every action cannot but contribute to them (and, in turn, re-performing the action would foster "recognition of and reflection on" their unavoidable operation). Yet the statement can also be read to imply locally constituted frames which (as "structures in their enactment") would be irreducibly contingent upon situated activity and thus point beyond any preexisting condition (structure, relation, or investment). To lift this ambiguity, the remainder of this chapter aims at explaining and exemplifying how EM, and practice-based video analysis in particular, may both re-embed the stated rationale and deliver upon its critical potential *in situ*, both with and against Fraser's position in and line of performance art.[6]

Machine Intelligence "On Stage": A Practice-Based Video Analysis

> The original question, "Can machines think?" I believe to be too meaningless to deserve discussion. (Turing, 1950, p. 442)

Over recent years, there has been a considerable STS interest in techno-scientific promises (Konrad et al., 2017). Just how, however, is any such promise actually made, delivered as a "performative utterance" (Austin, 1962) *in situ*, in and as its mundane expression? Just how might its mundane expression rely upon staged circumstances, especially when it comes to public announcements in and for the news media? And how might these circumstances alleviate Turing's initial skepticism, as encapsulated in the quote above and (luckily, I guess) still present today? To date, STS has not dwelled on these questions, perhaps both due to its late take-up of media studies (but see Badouard et al., 2016; Horst et al., 2017) and the only recent revival of "artificial intelligence" (see *Science*, 2015). In turn, this section examines three media announcements and how, in the form of video clips, they stage "machine intelligence" in one form or other.[7]

Announcement #1: "Console Creativity" (MoMA video games)

In August 2012, Paola Antonelli, then senior curator at the Department of Architecture and Design of New York's Museum of Modern Art (MoMA), announced the integration of video games to MoMA's collection of artworks, as a pioneering collector's gesture by the museum. Initially published as a blog entry (Antonelli, 2012), the announcement soon became the topic of a BBC feature, entitled "Console Creativity: Museums Display Video Games" (see Bressanin, 2013). A minute into the video, Antonelli, after having been introduced as "curator," takes the following stance: "Are video games art? I believe they are … and, eh, it's just a matter of time before the whole world believes that." How was this announcement of "console creativity" made and shown to be made, the announcement of video games as both museum-worthy and more broadly accepted techno-scientific artifacts? For a performative utterance to be effective, Austin notes in his seminal essay, it has to be stated under "ordinary circumstances" (Austin, 1962, pp. 25–26). How, then, does the video stage such circumstances? In the BBC feature, the announcement, before it is explicitly stated, is shown to be already realized in practice, at least partly. Indeed, the feature first shows a street view of MoMA, then a flat floor-filling artwork inside, and finally three male teenagers engaging in video gaming at the museum (one of them with his smartphone filming the other two at play). Against this visual background, the curator's announcement appears as a credible statement, all the more so as the visual background provides for MoMA's cultural leadership and social inclusiveness as an "ordinary circumstance," despite or precisely because of its gendered performance (for an educational account of MoMA's social inclusiveness, see Rosenberg, 2017).[8]

What may a re-enactment change or add to the outlined kind of video analysis? To have this question addressed, I asked students enrolled in a methods seminar in EM/ STS to re-enact the "console creativity" video. For the requested task, students were not given any technical assistance or special equipment as in a drama class or film course, but were simply asked to draw upon their practical skills and use their own digital devices (smartphones, video cameras, etc.). The students' re-enactment of a subsequent video episode, given their technical handicap with respect to professional standards, should afford us with a heuristic resource to examine its montage and the framings that it implies. Consider the excerpt in Figure 14.1:

<table>
<tr><th>Shot</th><th>Voice</th><th>Audio</th><th>Video</th></tr>
<tr><td>1
(0:08)</td><td>Narr

(Narr)

Narr</td><td>((street noise: siren and car))
video ga::mes are an a:rtform, in
the sense that |ro::lling a perfect
|((takes a puff))
#1

ci::garette is an artform.</td><td>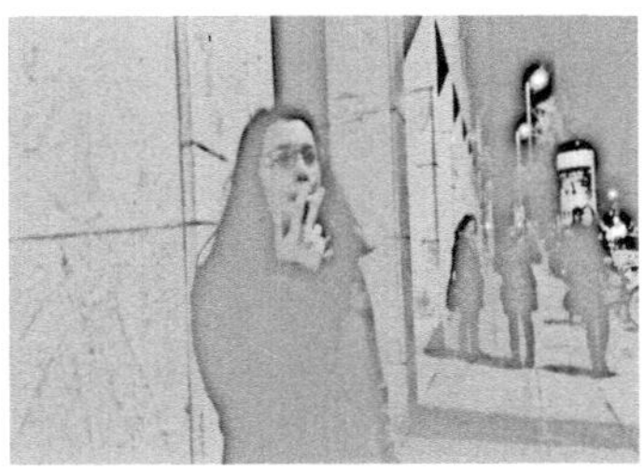
#1 (((Narr) takes a puff))</td></tr>
<tr><td>2
(0:11)</td><td>Narr

Narr

Narr
Narr</td><td>the question of whether or not
they are art |doesn't seem to me
|((spreads arms,
shakes head, shows
palms))
#2
a terribly interesting |question.
|((smiles))</td><td>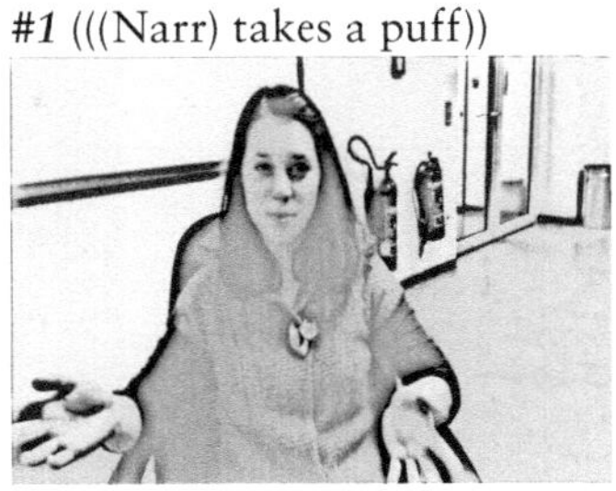
#2 ((Narr shows palms))</td></tr>
</table>

Figure 14.1 Re-enacted video episode ("Liam Gillick on console creativity").[9]

In the BBC feature, a second interlocutor is introduced after the curator's announcement, to elaborate his views on the artistic status of video games. Figure 14.1 shows the re-enactment of that episode. The re-enactment proves both problematic and tutorial, insofar as it manifestly fails to introduce that second interlocutor ("Liam Gillick, artist", as originally identified with a label in the second shot) and, in failing to do so, attunes us to the initial framing of his progressive introduction. First, he (not she) is shown smoking during daytime in an empty street with sunglasses on (not in the evening with glasses on and people walking home from work, *#1*). Second, he (again, not she) is shown surrounded by artworks inside a museum or gallery space (rather than in a university building corridor with fire extinguishers on the wall, #2). Taken together, the two successive shots of the original video thus show the interlocutor as an artist (smoking at daytime, hanging out in museums, and so on), before explicitly naming and labelling him as such (as "Liam Gillick, artist"). In so doing, the montage frames his commentary as entitled by virtue of his identity category and its "natural habitat", as this was the case with the previously identified curator. No further explanation is given. The re-enactment, in turn, makes explicit the clichéd framing in terms of which the ironic declaration "video games are an artform in the sense that rolling a perfect cigarette is an artform" (shot 1) may count as a credible, if ironic confirmation of the curator's announcement.[10]

Announcement #2: "Neuromorphic Computing" (Human Brain Project)

Neuromorphic computing, in contrast to video gaming, has not yet achieved the (arguable) status of collectable artwork. By and large, the idea of "modeling the computer after the brain" and, in so doing, revolutionizing computer engineering, medicine, and neuroscience has remained contested (e.g., *The Economist*, 2013; 2018). How was this promise initially made? And what kinds of framing – epistemic, aesthetic, institutional – would its making rely upon? In 2012, the press office of the Human Brain Project (HBP) released a promotional video as part of its bid for the European Union's 1-Billion Euro Flagship Initiative grant (see HBP, 2012). The promotional video addressed a broader audience in addition to the project selection committee. How, then, does the video's montage allow us, as a projected community, to "see through [our] cultural knowledge [and to] understand the filmic image and sequence ... in much the same way, and by reliance on the selfsame resources that we use to understand the perceptual world around us, a perceptual world of activity and interaction" (Jayyusi, 1988, p. 272)? Consider the excerpt from the original video in Figure 14.2:

Shot	*Voice*	*Audio*	*Video*
1 (4:30)	Narr	♫... **The |project will** |((researcher grasps computer)) ***#1*** **stimulate the development of** ...♫...	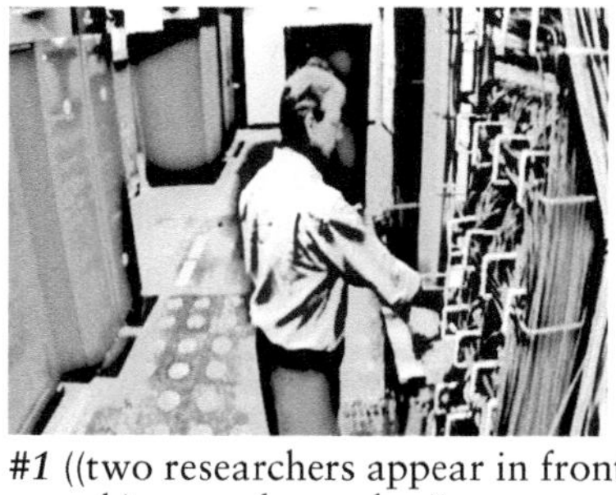 ***#1*** ((two researchers appear in front of (computing rack) – top-down shot))

Figure 14.2 Transcribed excerpt from promotional video ("HBP on neuromorphic computers").

<table>
<tr><th>Shot</th><th>Voice</th><th>Audio</th><th>Video</th></tr>
<tr><td>2
(4:33)</td><td>Narr</td><td>new supercomputers.
#2

…♫…</td><td>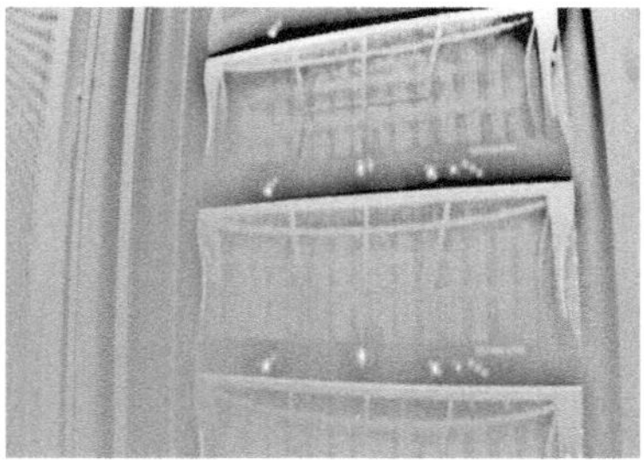
#2 ((black computing rack is shown – bottom-up shot))</td></tr>
<tr><td>3
(4:35)</td><td>Narr</td><td>By understanding how cognition works,
#3

it will help to design=

…♫…</td><td>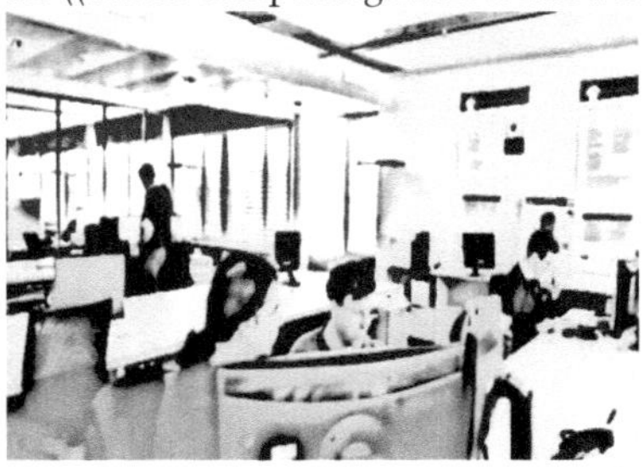
#3 ((open space office is shown – diagonal shot, from bottom left))</td></tr>
<tr><td>4
(4:39)</td><td>Narr</td><td>|radically new devices
|((typing))
|tac tac tac

called
#4

neuromorphic computers.

…♫…</td><td>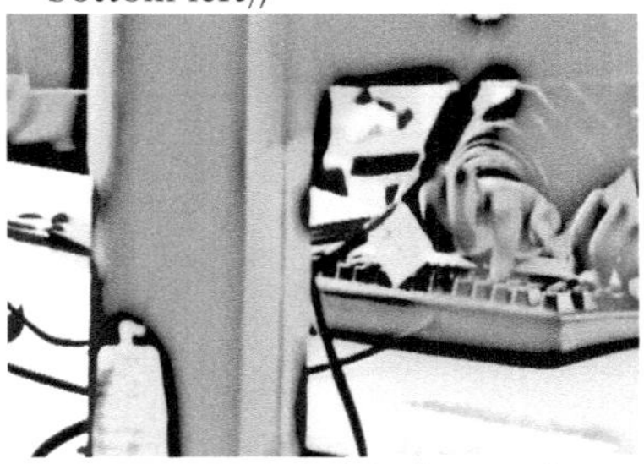
#4 ((typing hands, black, are shown close up))</td></tr>
<tr><td>5
(4:42)</td><td></td><td>…♫…
#5

…♫…</td><td rowspan="2">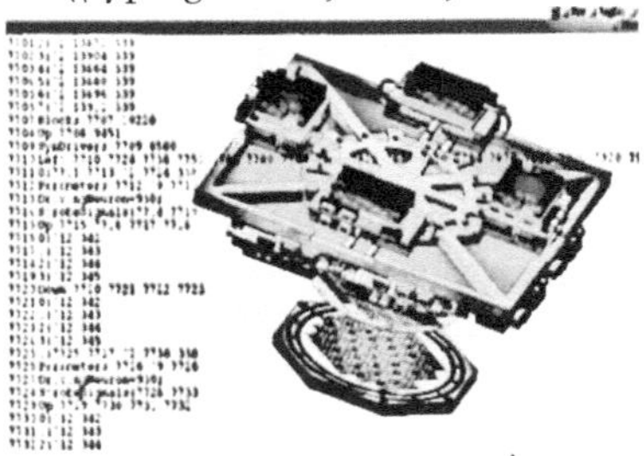
#5 ((pivoting digital 3-D model of “neuromorphic computer” is shown, against background of real time writing of computer code))</td></tr>
<tr><td></td><td>Narr</td><td>They will combine the power of=</td></tr>
<tr><td>6
(4:45)</td><td>Narr</td><td>=|microelectronics
|((watching computer screen))
#6

…♫…</td><td>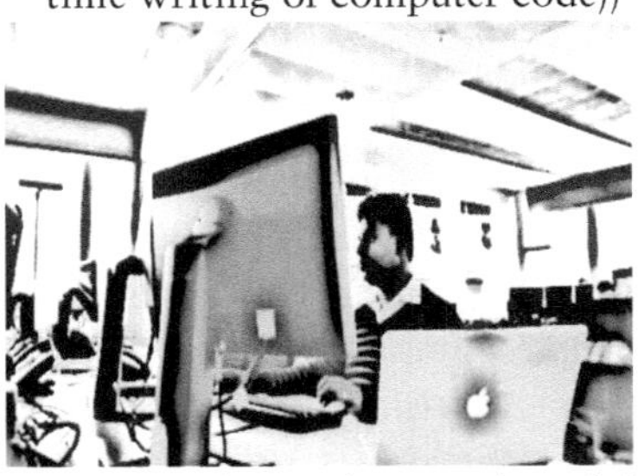
#6 ((researcher is shown – from left to right – Apple product placement on the right-hand side))</td></tr>
</table>

Figure 14.2 (Continued).

Shot	*Voice*	*Audio*	*Video*
7 (4:46)	Narr	**with the flexi|bility of** *#7* |((typing...)) |*tac tac tac...* ...♫... **human=**	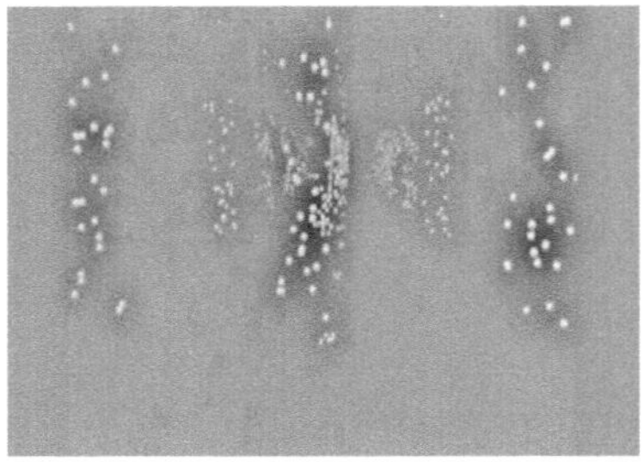 *#7* (((Dynamic simulated inside brain shot is shown)))
8 (4:47)	Narr	**=|intelligence'.** |((... typing))' |...*tac tac tac'*. *#8* '((open mouths)) ...♫...	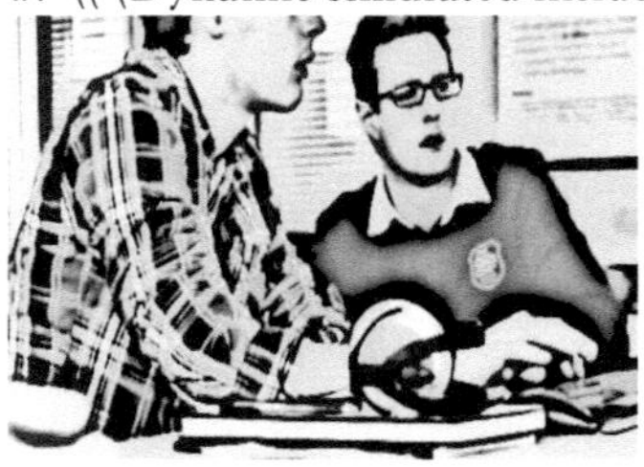 *#8* ((two young researchers are shown collaborating at computer))

Figure 14.2 (Continued).

The transcribed excerpt allows us to examine how the promise of "neuromorphic computers" is actually delivered, with particular reference to the interactional resources that its delivery relies upon. Notice that, similarly to our previous example, the promise in question is not only progressively delivered ("Narr", audio column), but also shown to be already in the process of being fulfilled (video column). Moreover, the successive shots, organized in terms of as many "say-shows" (Garfinkel, 2002, p. 177), articulate a particular "news delivery sequence" (Maynard 1997). From ordinary conversation, the sequence borrows its constitutive parts: an announcement, followed by its response, an elaboration, and a final assessment (ibid., p. 97). Yet the parts are not distributed as reciprocating turns at talk, as in conversation, but across the successive "say-shows" that compose the video episode. Accordingly, the "announcement" is first stated ("the project will stimulate the development of new supercomputers" [shots 1 & 2, audio column]), for the viewer then to be enabled to reach a positive "response" (as a bottom-up shot casts the "new superconductors" as something to be looked up to [shot 2, #2]). The subsequent "elaboration," in turn, suggests that the "understanding [of] how cognition works" (shot 3, audio) has already progressed so swiftly (*tac tac tac* [shot 4 onwards]) that, indeed, not only first "neuromorphic computers" (shot 4, audio) are already being designed (shots 4 & 5, *#4* & *#5*), but also that a longstanding opposition, between "microelectronics" and "human intelligence," is in the process of being overcome (shots 5–8). Finally, the "assessment" of progress is left to the researchers shown in the video (shot 8), as they stare in awe at their computer screen (*#8*), presumably showing their first results, which remain withheld from the viewer, however.[11]

In the "perceptual world of activity and interaction," to use Jayyusi's phrase (1988, p. 272), conversationalists "shape each component [of a news delivery sequence] according to a *myriad of contingences*," as Maynard points out (1997, p. 98; emphasis added). The re-enactment of the video announcement of neuromorphic computing allows us to tease out three such critical contingencies and thus to turn them into "tutorial problems" (Garfinkel, 2002, ch. 4). First, and as I had EM/STS students re-enact the opening sequence of the promotional video too, I noticed that a brain/machine analogy was posited from the outset, casting "the human brain [as] the most complex machine that we know of, and the most mysterious one" (HBP, 2012), thus pre-empting subsequent doubts with respect to the conceptual cogency or empirical possibility of combining "microelectronics" and "human intelligence." Second, I actually happened not only to visit the Swiss National Supercomputing Centre (CSCS) in Lugano, powering part of the HBP, but did also attempt to re-enact *in situ* the researcher's grasp as shown above (Figure 14.2, shot 1, *#1*). In so doing, I not only noticed the persistent noise of the cooling system at the CSCS ("Shhhshhhhshhhhhsh-hhh... ", consistently edited out of and covered with music, ...♫..., in the promotional video, shots 1–8) but also that neither my re-enacted grasp, nor the original grasp (shot 1, *#1*) or, for that matter, any of the subsequent gestures and manipulations shown in the promotional video (cf. shots 3, 4, 6, 8, *#3*, *#4*, *#6*, *#8*) appear to make a clear and clearly identifiable contribution to "neuromorphic computing" *per se*. Finally, the diverse background of the EM/STS student cohort (of Asian, African, and European origin, both female and male) throws into relief the restricted "old boy network" from which the HBP spokespersons were drawn (certainly, and understandably, not including critical voices, echoing either Ryle's [1949] or anticipating Penny's [2017] arguments, for example).[12]

Announcement #3: "Machine Learning" (Google DeepMind's AlphaGo)

Our last example of a media announcement of machine intelligence took a textual form, as it was posted on the company website of DeepMind, a Google-owned and London-based company specializing in "machine learning" applications. The announcement is based upon a best-of-five exhibition match of the game of go, livestreamed from Seoul, South Korea, in early March 2016. The exhibition match pitched the company's go program, AlphaGo, against a top go professional, Lee Sedol, sometimes nicknamed "the Federer of go." The media announcement reads as follows:

> During the games, AlphaGo played a handful of highly inventive winning moves, several of which – including move 37 in game two – were so surprising they overturned hundreds of years of received wisdom, and have since been examined extensively by players of all levels. AlphaGo somehow taught the world completely new knowledge. (DeepMind, 2018)

In the second game of the exhibition match, the aforementioned move ("move 37") came as a big surprise to the locally present game commentators, indeed. It manifestly constituted a surprise move to them all the more so as one of them (Michael Redmond, present and presented as the strongest professional go player in the West) had just explained that it would be inefficient to play in the very area where the move was eventually placed (by Aja Huang, Google DeepMind's lead programmer, embodying and enacting AlphaGo during the exhibition match (Figure 14.3, lines 1–3, 5–6 versus line 4, #1).

01 MR **it's not gonna spread out into the center, because of these black stones in the**
02 **way, and so. the value here is relatively slo- small, because of it doesn't.**
03 **spread out.**
04 AG **and |also because these stones.**
|#1 ((plays out move 37))

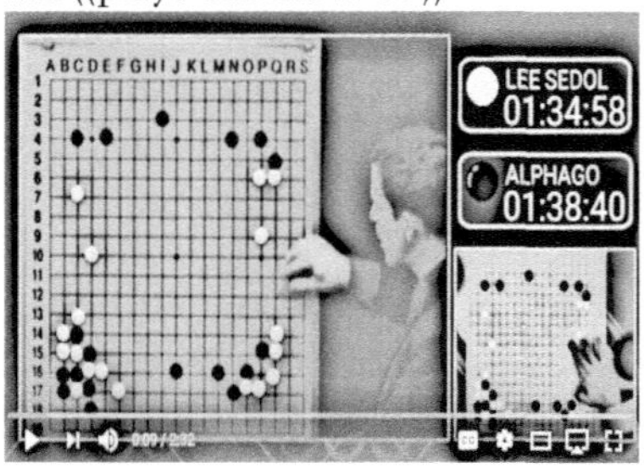

((*left-hand side*: front shot of commentator room, showing MR commenting with help of magnetic board))

((*right-hand side*: simultaneous top-down shot on go board in game room, with Aja Huang's hand playing AG's move 37))

05 MR #1 ((MG comments, AG plays move 37))
06 **if black chose to. play a lot of moves on the right side. he would not have the added value of having eh, say, an attack**

Figure 14.3 Move 37 (as played by AlphaGo [AG]) and contrary commentary (as made by Michael Redmond [MR]).

How was the exhibition match staged in the first place, not only for surprise moves to become possible but also for its successive games to appear as part of a regular "game of go"? A first contingency to be dealt with was indeed the very intelligibility of the staged encounter. The contingency was perhaps best formulated by a post-game commentary in the *Neue Zürcher Zeitung*: "no offense taken, dear algorithm developers – but does it make much sense to simply compare Usain Bolt with a new Formula-1-Ferrari?" (Kaeser 2016). To accommodate this contingency, the encounter was staged as a "symmetry spectacle," showing two opponent players in the game room (Lee Sedol, the go professional, and AlphaGo, represented by Aja Huang) and two commentators in another room, the commentator room (Michael Redmond and Chris Garlock, managing editor of the *American Go E-Journal*). This split-screen arrangement provided for the scenic intelligibility of the livestreamed game of go two as part of a regular exhibition match (see Fig. 14.3, which zooms in on MR). In turn, the required infrastructure of computer networks to operate AlphaGo was not shown, but backgrounded (including Google's so-called "tensor processing units").[13]

As we re-enacted "move 37," I was made aware of further contingencies that the initial screening had taken into account, for them to be accommodated and AlphaGo's machine intelligence to appear (Figure 14.4).

First, the top-down shot of the game opposing Lee Sedol and AlphaGo (Figure 14.4a) puts the viewer in the position of AlphaGo, via a "camera subjectiva" shot that has AlphaGo's hand (embodied by the unnamed Aja Huang) appear as the virtual prolongation of the viewer's body (so that he or she indeed can see him- or herself as playing AlphaGo's "move 37" and, thereby, virtually experience "machine intelligence"). Tech-savvy artist Hunter Longe first noticed this during our collaborative re-enactment (hence my use of "we" above). Second, his re-enactment of "move 37" (as shown in Figure 14.4b) led me to notice that the original move did *not* exhibit

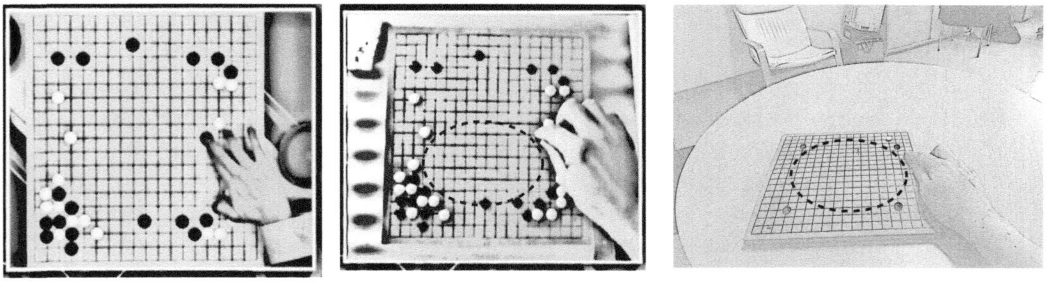

Figure 14.4 Move 37, as initially played (14.4a), re-enacted (14.4b), and compared with re-enacted opening move (14.4c).

"completely new [go] knowledge" (as claimed on DeepMind's website), but displayed conventional go skill in at least two respects. On one hand, the gesture with which Aja played AlphaGo's move 37 should be re-examined (see Figure 14.4a). Indeed, he manifestly not only enacted a conventional *te-tsuki*, as the physical gesture of playing go moves is named in Japanese (where the go stone is held between the index and middle finger as shown), but he also did it so confidently that AlphaGo's black stone made a quick left–right move on the go board, a horizontal slip due to the finger pressure with which it was played out (in contrast to Hunter's clumsy gesture [Figure 14.4b]). On the other hand, it should be noticed that move 37 exhibits a classic principle of "power play" in go (Figure 14.4b), namely the principle according to which one should stake out one's territorial claims over the whole board (as indicated by the area inside the dotted line, potentially surrounded by AlphaGo's black stones), and not only in local terms of smaller configurations (Kajiwara, 1979). A re-enactment of the opening in a prior teaching game reminded me of this principle (Figure 14.4c), a principle that Michael Redmond was also seeing to be at play in the exhibition match, at least as far as "move 37 in game two" was concerned (as he put it subsequently to the excerpted episode, and by using Kajiwara's term, "the *direction of play* does not seem unreasonable").[14]

Discussion

> Procedurally, it is my preference to start with familiar scenes and ask what can be done to make trouble. (Garfinkel, 1967, p. 37)

Defended as a postmodernist gesture in the early 1990s, this Dada-esque impulse of EM may produce "fundamental disturbances of established, familiar, and corporately approved ways of making sense" (Lynch & Bogen, 1991, p. 269). Taking its cue from Garfinkel's EM, the present chapter echoed a subversive move of this kind to emphasize the *heuristic* character of deliberately troubled re-enactments (that is, to figure out what "ways of making sense" may be involved in the first place). In so doing, the chapter also may have resonated with historical works of performance art, "works [that] used abbreviated scores and were under-rehearsed precisely in order to elicit variation, risk, and unpredictability" (Bishop, 2017, p. 40). In addition to Futurist and Dadaist performances, Bishop's discussion refers to actionist happenings of the 1960s and Fluxus' "playful (and occasionally violent) subversions of the classic music recital – smashing violins, pianos, etc." (ibid.). These, on Bishop's account (but see also Stewart-Halevy 2017), exemplify

> visual performance art [that sets] itself in opposition to the conventional trappings of professional theatre (costumes, characters, narrative, scripts, and rehearsals) in favor of everyday clothing, actions, and situations. ... Events did not take place in a purported elsewhere (signaled by a proscenium arch and set design), but in the actual space-time of the viewer. These contingencies of location and audience were an important part of performance art's claims to realism, authenticity, and democratic anti-virtuosity. (Bishop, 2017, p. 40)

Over the last three decades, Andrea Fraser in turn has contributed to the institutional critique of contemporary art (e.g., MACBA, 2016) by appropriating, enacting and re-enacting many of its conventions, participant roles and characteristic moments (for example, the museum visit). Accordingly, her performance art has highlighted the hierarchical structures and specific closure of the arts field produced and reproduced through these conventions, on the one hand, and argued for more engaged and articulated, if not liberating self-criticism, on the other (e.g., Fraser, 2009). In contrast to Dada-esque tactics, Fraser's performances indulge in the ostensibly *over-rehearsed* enactment of fads and foibles of contemporary art, including the awkwardly over-informed gallery tour by Jane Castleton, a fictitious docent largely, if not absurdly, oblivious of her public (Fraser, 1989). However, and although Fraser's *reductio ad absurdum* typically challenges established art institutions, ranging from the museum and gallery space to the studio, her artwork-cum-critique does not seem to imagine alternative formats, let alone invent new kinds of institutions. To the contrary, on her account, "we are trapped in our field" (Fraser, 2009, p. 415).[15]

The outlined argument against Fraser's institutional critique is not without surface appeal. What the argument seems to overlook though is her critique's productive ambivalence (as hinted at in the first section above). Indeed, if Fraser's interest is in "structures in their enactment," rather than structures, institutions or conventions *per se*, then this constitutes an apt reminder of their locally contingent constitution, involving myriad "vulgar enabling practices" (Button & Sharrock, 1995). Truly, this reminder may have been articulated more aptly in and through Fraser's performances and ironic set pieces *ad hoc* than their socio-theoretical sources (for example, Pierre Bourdieu's field theory) or psychoanalytic principles. Practice-based video analysis at least afforded us the opportunity to invite this reflexively praxeological, rather than a comprehensively structuralist, reading of Fraser's institutional critique *qua* filmed intervention. That is to say, Fraser's performances, precisely because of their over-rehearsed character and absurd stringency, constitute an invitation to challenge, criticize, and change routine practice and its instituted frames (and be it only the over-extended use of the notions of "performance" and "performative" in art discourse). Whether, why, and how this invitation is to be accepted, to my mind, is not for Fraser to answer. The same holds for practice-based video analysis.[16]

This chapter, by and large, drew upon the latter approach to probe media announcements of machine intelligence, a long-standing promise, marketing fetish, and multi-faceted field that Fraser's performance art, as yet, has not appropriated or engaged with. This lack of engagement begs the perennial question of "technology and society," be it in terms of its aesthetic dimension, economic implications, or epistemic politics, including the de- and re-skilling of artistic practice, its "equipped rhetoric" (Rosental, 2017, p. 2), as well as the reconfiguration of work at large, digitally or otherwise (cf. Bishop, 2017; Doing, 2016; Penny, 2017). In probing three scenic performances of technological determinism, this chapter contributed to its "infrastructural critique" (cf. Holert, 2016), if only by demonstrating how smooth technical operations were enacted through, if not as tacit audiovisual montages.

Appendix: Transcription Conventions

=	latching, no discernible interval between adjacent utterances or activities
she-	cut-off
he: ::	elongation (successive colons indicating increased elongation)
.	falling intonation
,	"continuing" intonation
(go ahead)	uncertain hearing or seeing
((does))	description, comment
| '	comment on non-verbal activity, one sign per participant, if there is a verbal line, marked on the verbal line and again on the comment line E.g.: **in the sense that |ro::lling a perfect ci::garette is an artform.** |(((Narr) takes a puff))
#1	indication of screenshot placement in the transcribed activity
♫…♫…	onset and continuation of background music
tac tac tac	typing noise

Acknowledgments

This chapter has benefitted from remarks and observations by Alain Bovet, Luca Greco, Hunter Longe, Federica Martini, Claude Rosental, the editors of this book, as well as several EM/STS students and the participants of the panel at 4S/EASST 2016 in Barcelona at which the chapter was initially presented. Whilst their remarks and observations helped me to improve the chapter, I solely remain responsible for its final version.

Notes

1 For related discussions of re-enactment in historical research, design, and contemporary art, see *inter alia* Campos & Donin, 2016; Cangiano et al., 2018; Fors et al., 2016; *Marges*, 2013. For a recent video ethnography, see Pink & Leder Mackley (2014).

2 For a previous position statement advocating the indicated three-step approach, see Sormani, 2016.

3 The tricky character of this intersection may be glossed in terms of Duchamp's discontentment: "Duchamp was not content with the liberties available in art when viewed as opposed or complementary to science. By blending scientific and aesthetic logic he rejected both science's veneration of objectivity and truth and art's veneration of the individual" (Bippus, 2013, p. 127).

4 In the opening pages of *Scientific Practice and Ordinary Action*, the landmark volume that charts the tangled relationship between EM and STS up to the early 1990s, Michael Lynch commented upon a structurally similar paradox, namely the paradoxical "irrelevance of professional sociology [to] transdisciplinary critical discourse" (Lynch 1993, p. xiv). Indeed, scholars engaging in critical discourse across the humanities at the time, although

they engaged with social practices, textual criticism, and traditional topics of sociology ("race," "class," "gender," "power," etc.), would typically shun the canonical outlets, empiricist arguments, and "scientistic style that dominate[d] American sociology" (ibid.).

5 As Fraser contends, "even our capacity to use 'performative' to reflect specifically on what we may be doing with words and other non-actions has been lost in artistic usage [and art discourse]" (2014, p. 124).

6 The EM orientation of the suggested (mis-)reading is nicely captured by Lucy Suchman's phrase: "lived practice inevitably exceeds the enframing moves of its own procedures of order production" (2007, p. 193).

7 On Wittgenstein's critique of Turing's subsequent epistemology in favour of "machine intelligence," see Shanker, 1987.

8 The "machine intelligence" of video games as such is presupposed, where this presupposition then makes it possible to ask (and answer) the question whether video games are to be considered "art." For a recent analysis that examines video gaming as a practical activity, rather than a presupposed intelligence, see Reeves et al., 2016.

9 In the transcript, "Narr" stands for "narrator." For the complete transcription conventions, see Appendix, and, for the original video episode, see Bressanin (2013).

10 In other words, the video stages the circumstances shown (the empty street during daytime, a museum space, etc.) so as to provide for the identity categories' and related statements' plausibility (*qua* "established curator" or quasi "marginal artist"). Note also, in this respect, the gendered appearance of the artist and curator, as well as the subtle shift from video games identified as "art" and/or "design" (by the curator) to them being identified as an "artform" (by the artist). A recent plea for categorization analysis along these lines is to be found in Fitzgerald & Housley, 2015.

11 As in ordinary conversation, a preannouncement precedes the analyzed "news delivery sequence" (see Maynard, 1997, pp. 95–96). The preannouncement, in the video, is stated as follows: "the simulation of a complete human brain will require supercomputers that are a thousand times more powerful than today" (HBP, 2012). As in conversation, it is "designed to handle a central contingency in the development of … news, which is a recipient's prior knowledge of the occurrence to be reported" (Maynard, 1997, p. 95).

12 As in the "console creativity" episode, interlocutors and spokespersons, all male and mostly senior, are identified in official terms (e.g., as director) and shown under circumstances (e.g., a laboratory setting, a seminar room, a university building) which make their statements credible and their comments plausible (where the identity categories are elaborated by the circumstances shown, including their musical post-production, and vice-versa).

13 Incidentally, both the machine-learning history of AlphaGo (having been exposed to and improved by millions of games) and the professional identity of Aja Huang (as lead programmer) were backgrounded. For a recent documentary movie on AlphaGo backstage, see Krieg & Kohs, 2017.

14 Interestingly, DeepMind explains AlphaGo's "highly inventive winning moves" in Kajiwara's spirit too: "Although Go is a game of territory, most decisive battles hinge on the balance of power between groups, and AlphaGo excels in shaping this balance" (DeepMind, 2018).

15 Fraser's appropriationist art ruminates the "conventional trappings of contemporary art" (to borrow Bishop's turn of phrase) to a point where stating one's opposition to them becomes futile (that is, *if* one accepts Fraser's "closure phantasms") (Wilson, 2017, p. 115). For this line of criticism, see e.g. Raunig, 2009; Wilson, 2017; Von Hantelmann, 2010, pp. 167–193.

16 Instead of gesturing at a "pure critique" (e.g., Von Hantelmann, 2010, p. 181), both approaches stage "controlled irony" as a performative contradiction (Parcero Oubiña, 2006), thus inviting a "second look" at established ways of working, aesthetic or epistemic, if not instigating institutional change (in this vein, see also Medina, 2016; Rebentisch, 2016).

References

Abramović, M. (2016). *Walk through Walls: A Memoir.* New York: Crown Archetype.

Adamson, G., & Bryan-Wilson, J. (2016). *Art in the Making: Artists and their Materials from the Studio to Crowdsourcing.* London: Thames & Hudson.

Antonelli, P. (2012). Video Games: 14 in the Collection, for Starters. Retrieved from www.moma.org/explore/inside_out/2012/11/29/video-games-14-in-the-collection-for-starters.
Austin, J. L. (1962). *How to Do Things with Words*. Cambridge, MA: Harvard University Press.
Badouard, R., Mattozzi, A., Sire, G., Mabi, C., Schubert, C., & Sørensen, E. (2016). STS and Media Studies: Alternative Paths in Different Countries, *Tecnoscienza: Italian Journal of Science and Technology Studies*, 7(1), 109–128.
Bippus, E. (2013). Artistic Experiments as Research. In M. Schwab (Ed.), *Experimental Systems: Future Knowledge in Artistic Research* (pp. 121–134). Leuven: Leuven University Press.
Bishop, C. (2017). Performance Art vs. Dance: Professionalism, De-Skilling, and Linguistic Virtuosity. In C. Costinas & A. Janevski (Eds.), *Is the Living Body the Last Thing Alive? The New Performance Turn, Its Histories and Its Institutions* (pp. 40–45). Berlin: Sternberg Press.
Borgdorff, H. (2012). *The Conflict of the Faculties: Perspectives on Artistic Research and Academia*. Amsterdam: Leiden University Press.
Bressanin, A. (Producer) (2013). Console Creativity: Museums Display Video Games. *BBC*. Retrieved from https://www.bbc.com/news/av/magazine-21661690/console-creativity-museums-display-video-games.
Broth, M., Laurier, E., & Mondada, L. (Eds.) (2014). *Studies of Video Practices: Video at Work*. London and New York: Routledge.
Butler, J. (1990). *Gender Trouble: Feminism and the Subversion of Identity*. London: Routledge.
Butler, J. (2015). *Notes Toward a Performative Theory of Assembly*. Cambridge, MA: Harvard University Press.
Button, G., & Sharrock, W. (1995). The Mundane Work of Writing and Reading Computer Programs. In P. Ten Have & G. Psathas (Eds.), *Situated Order: Studies in the Social Organization of Talk and Embodied Activities* (pp. 231–258). Washington DC: University Press of America.
Campos, R., & Donin, N. (2016). Réactiver des situations passées? Du *re-enactment* à l'histoire pragmatique. In F. Chateauraynaud & Y. Cohen (Eds.), *Histoires pragmatiques* (pp. 247–288). Raisons Pratiques, 25, Paris: Ed. de l'EHESS.
Cangiano, S., Fornari, D., & Seratoni, A. (2018). Re-Programmed Art: Overcoming Obsolescence and Opening up Programmed and Kinetic Artworks. In P. Sormani, G. Carbone, & P. Gisler (Eds.), *Practicing Art/Science: Experiments in an Emerging Field* (pp. 197–215). London and New York: Routledge.
Corner College (2017). What's Wrong with Performance Art? A Trans-local Sympodium. Retrieved from http://materials.corner-college.com/2017/201710/20171027-29-sympodium-what-s-wrong-with-performance-art.html.
Costinas, C., & Janevski, A. (Eds.) (2017). *Is the Living Body the Last Thing Alive? The New Performance Turn, Its Histories and Its Institutions*. Berlin: Sternberg Press.
DeepMind (2018). The Story of AlphaGo So Far. Retrieved from https://deepmind.com/research/alphago.
Doing, P. (2016). Epistemic Politics at Work: National Policy, an Upstate New York Synchrotron, and the Rise of Protein Christallography, *Sociology of the Sciences Yearbook, 29*, 137–152.
Fitzgerald, R., & Housley, W. (Eds.) (2015). *Advances in Membership Categorization Analysis*. London: Sage.
Fors, H., Principe, L. M., & Sibum, H. O. (2016). From the Library to the Laboratory and Back Again: Experiment as a Tool for Historians of Science, *Ambix*, *63*(2), 85–97.
Fraser, A. (Producer and Director) (1989). *Museum Highlights: A Gallery Talk* [video, 29 min loop]. https://www.artic.edu/artworks/193356/museum-highlights-a-gallery-talk?artist_ids=Andrea%20Fraser.

Fraser, A. (2009). From the Critique of Institutions to an Institution of Critique [2005]. In A. Alberro & B. Stimson (Eds.), *Institutional Critique: An Anthology of Artists' Writings* (pp. 408–425). Cambridge, MA: MIT Press.

Fraser, A. (2014). Enactment or Performance. In C. Dertnig & F. Thun-Hohenstein (Eds.), *Performing the Sentence: Research and Teaching in Performative Fine Arts* (pp. 122–127). Vienna and Berlin: Akademie der bildenden Künste and Sternberg Press.

Garfinkel, H. (1967). *Studies in Ethnomethodology*. Englewood Cliffs, NJ: Prentice-Hall.

Garfinkel, H. (2002). *Ethnomethodology's Program: Working Out Durkheim's Aphorism*, ed A. Rawls. Lanham, MD: Rowman & Littlefield.

Garner, D. (2016). A Memoir of Masochism and Pretension. *New York Times*, international ed. (5–6 Nov.), pp. 1–2.

Gergen, M. M., & Gergen, K. J. (2012). *Playing with Purpose: Adventures in Performative Social Science*. Walnut Creek, CA: Left Coast Press.

Goldberg, R. (2011). *Performance Art. From Futurism to the Present* [1979], 3rd ed. London: Thames & Hudson.

Goldberg, R., Curiger, B., Burckhardt, J., & Columbus, N. (2014). Performance Record: Rose-Lee Goldberg, Bice Curiger, and Jacqueline Burckhardt in Conversation with Nikki Columbus, *Parkett*, *94*, 6–29.

HBP (2012). *The Human Brain Project: Video Overview* [video]. Retrieved from https://www.youtube.com/watch?v=JqMpGrM5ECo.

Herzig, R. (2004). On Performance, Productivity, and Vocabularies of Motive in Recent Studies of Science, *Feminist Theory*, *5*(2), 127–147.

Holert, T. (2016). A Politics of Knowledge in Contemporary Art? *Performance Research: A Journal of the Performing Arts*, *21*(6), 57–62.

Horst, M., Davies, S. R., & Irwin, A. (2017). Reframing Science Communication. In U. Felt, R. Fouché, C. A. Miller, & L. Smith-Doerr (Eds.), *Handbook in Science and Technology Studies*. 4th ed. (pp. 881–908). Cambridge, MA: MIT Press.

Jayyusi, L. (1988). Toward a Socio-logic of the Film Text, *Semiotica*, *68*(3–4), 271–296.

Kaeser, E. (2016). Ein unfairer Vergleich – Gastkommentar. *Neue Zürcher Zeitung* (8 Apr.). Retrieved from https://www.nzz.ch/meinung/kommentare/alpha-gos-scheinsieg-ueber-die-menschliche-intelligenz-ein-unfairer-vergleich-ld.12145.

Kajiwara, T. (1979). *The Direction of Play*. Tokyo: Ishi Press.

Konrad, K., Van Lente, H., Groves, C., & Selin, C. (2017). Performing and Governing the Future in Science and Technology. In U. Felt, R. Fouché, C. A. Miller, & L. Smith-Doerr (Eds.), *Handbook in Science and Technology Studies*, 4th ed. (pp. 465–493). Cambridge, MA: MIT Press.

Krieg, G. (Producer) & Kohs, G. (Director) (2017). *AlphaGo* [documentary]. Moxie Pictures.

Lynch, M. (1993). *Scientific Practice and Ordinary Action: Ethnomethodology and Social Studies of Science*. Cambridge: Cambridge University Press.

Lynch, M., & Bogen, D. (1991). In Defense of Dada-Driven Analysis, *Sociological Theory*, *9*(2), 269–276.

MACBA (2016). *Andrea Fraser: L'1%, C'est Moi, Exhibition Website, Barcelona Museum of Contemporary Art*. Retrieved from https://www.macba.cat/en/exhibition-andrea-fraser.

Marges (2013). *Remake, reprise, répétition* [special issue], *17*.

Maynard, D. W. (1997). The News Delivery Sequence: Bad News and Good News in Conversational Interaction, *Research on Language and Social Interaction*, *30*, 93–130.

Medina, C. (2016). Fragments of Autonomy. In E.A. Romero (Ed.), *Andrea Fraser: L'1%, C'est Moi, Exhibition Catalogue* (pp. 16–24). Barcelona and Mexico City: Barcelona Museum of Contemporary Art and MUAC, Museo Universitario Arte Contemporáneo, UNAM. Retrieved from https://www.macba.cat/en/exhibition-andrea-fraser.

Parcero Oubiña, O. (2006). "Controlled Irony" ... Are You Serious? Reading Kierkegaard's Irony Ironically. In N. J. Cappelorn & H. Dreuser (Eds.), *Kierkegaard Studies* (pp. 241–260). Berlin and New York: Walter de Gruyter.

Penny, S. (2017). *Making Sense: Cognition, Computing, Art, and Embodiment.* Cambridge, MA: MIT Press.

Pink, S., & Leder Mackley, K. (2014) "Re-enactment methodologies for everyday life research: art therapy insights for video ethnography", *Visual Studies*, *29*, 146–154.

Raunig, G. (2009). Instituent Practices: Fleeing, Instituting, Transforming. In G. Raunig & G. Ray (Eds.), *Art and Contemporary Critical Practice: Reinventing Institutional Critique* (pp. 3–12). London: MayFly Books.

Rebentisch, J. (2013). *Theorien der Gegenwartskunst. Eine Einführung.* Hamburg: Junius Verlag.

Rebentisch, J. (2016). *The Art of Freedom: On the Dialectics of Democratic Existence.* Cambridge: Polity Press.

Reeves, S., Greiffenhagen, C., & Laurier, E. (2016). Video Gaming as Practical Accomplishment: Ethnomethodology, Conversation Analysis, and Play, *TopiCS: Topics in Cognitive Science*, *9*(2), 308–342.

Rosenberg, F. (2017). What Does It Mean To Be An Accessible Museum, *MoMA Stories.* Retrieved from https://stories.moma.org/what-does-it-mean-to-be-an-accessible-museum-9e9708254dc9.

Rosental, C. (2013). Toward a Sociology of Public Demonstrations, *Sociological Theory*, *31*(4), 343–365.

Rosental, C. (2017). Montrer de quoi demain sera fait. *CEMS Occasional Paper*, *35*. Paris: CEMS, EHESS. Retrieved from http://cems.ehess.fr/index.php?3627.

Ryle, G. (1949). *The Concept of Mind.* Chicago: University of Chicago Press.

Salter, C., Burri, R.V., & Dumit, J. (2017). Art, Design, and Performance. In U. Felt, R. Fouché, C.A. Miller, & L. Smith-Doerr (Eds.), *Handbook in Science and Technology Studies*, 4th ed. (pp. 139–168). Cambridge, MA: MIT Press.

Schechner, R. (2002). *Performance Studies: An Introduction.* London and New York: Routledge.

Science (2015). *Artificial Intelligence* [special issue], *349*(6245), 248–278.

Shanker, S.G. (1987). Wittgenstein versus Turing on the Nature of Church's Thesis, *Notre Dame Journal of Formal Logic*, *28*(4), 615–649.

Shapin, S. (1995). Cordelia's Love: Credibility and the Social Studies of Science, *Perspectives on Science, 3*, 255–275.

Sormani, P. (2016). Practice-Based Video Analysis: A Position Statement, *SocietàMutamentoPolitica*, *7*, 103–120. Retrieved from http://www.fupress.net/index.php/smp/article/view/19698.

Sormani, P., Carbone, G., & Gisler, P. (Eds.) (2018). *Practicing Art/Science: Experiments in an Emerging Field.* London and New York: Routledge.

Sormani, P., Alač, M., Bovet, A., & Greiffenhagen, C. (2017). Ethnomethodology, Video Analysis, and STS. In U. Felt, R. Fouché, C.A. Miller, & L. Smith-Doerr (Eds.), *Handbook in Science and Technology Studies*, 4th ed. (pp. 113–137). Cambridge, MA: MIT Press.

Stewart-Halevy, J. (2017). The Inductive Turn in Conceptual Art: Pragmatics in the *0–9* Circle, *Grey Room*, *68*, 60–93.

Suchman, L. (2007). *Human-Machine Reconfigurations. Plans and Situated Actions*, 2nd ed. Cambridge: Cambridge University Press.

The Economist (2013). Neuromorphic Computing: The Machine of a New Soul (3 Aug,). Retrieved from https://www.economist.com/science-and-technology/2013/08/03/the-machine-of-a-new-soul.

The Economist (2018). AI, EU, Go: How Europe can Improve the Development of AI (22 Sept). Retrieved from https://www.economist.com/leaders/2018/09/22/how-europe-can-improve-the-development-of-ai.

Turing, A. (1950). Computing Machinery and Intelligence, *Mind, 49*, 433–460.

Turner, V. (1988). *The Anthropology of Performance.* Cambridge, MA: PAJ Publications.

Von Hantelmann, D. (2010). *How to Do Things with Art* [2007]. Zurich and Dijon: JRP Ringier and Les Presses du Réel.
Wilson, M. (2017). Institution and the Political Community with the Dead. In P. O'Neill, L. Steeds, & M. Wilson (Eds.), *How Institutions Think: Between Contemporary Art and Curatorial Discourse* (pp. 114–121). Cambridge, MA: MIT Press.
Wilson, M., & Van Ruiten, S. (Eds.) (2013). *SHARE Handbook for Artistic Research Education*. Amsterdam: ELIA.

Index

Note: *italic* page numbers refer to figures.

For Product Safety Concerns and Information please contact our EU representative GPSR@taylorandfrancis.com Taylor & Francis Verlag GmbH, Kaufingerstraße 24, 80331 München, Germany